THE ROMAN ORIGINS OF OUR CALENDAR

VAN L. JOHNSON

Tufts University
Medford, Mass.

LXXXIII
AMERICAN CLASSICAL LEAGUE
1974

CONTENTS

Page

I. Preface 5

II. Introduction with two illustrative diagrams 7

III. *Fasti Romani:* a reconstructed Roman Calendar 20

IV. Notes on the Roman calendar 33

V. Glossary of gods 75

VI. Bibliography 79

Fragment of the calendar of Praeneste for the beginning of March, carved in the first century of our era when the month of March had been firmly established as the third month of the year. Fragments of this calendar have been coming to light since the fifteenth century. The first column of letters indicates the eight days of the Roman week, lettered A to H. The column of Roman numerals designates days of the month as so many days before the Nones or Ides. The Kalends and Nones are abbreviated K and NON. The second column of letters gives the legal status of the days involved and to the right of this column are comments, now very fragmentary, on the festivals or dedications which occur. The author of these comments was probably Verrius Flaccus, a great scholar of the Augustan Age and a native of Praeneste (modern Palestrina). Photograph courtesy of Fototeca di architettura e topografia dell' Italia antica.

Reprinted from *Archaeology*, Vol. 21, No. 1 (January 1968) page 16.

I. PREFACE

One mark of an educated person is to take nothing for granted. This is not easy for most of us become so familiar with things around us that we assume their permanent existence and fail to examine the interesting facts which made them possible. The calendar is a case in point: we use it every day of our lives and seldom stop to think about the long process of circumstances which produced the convenience, or maybe the inconvenience, of this common mode of time-reckoning. In fact, the best argument against calendar reform, intermittently broached and sometimes for good cause, is that it might destroy too many of these intriguing connections with the past. The calendar, as we have it, is remarkable evidence of the persistence and validity of ancient tradition. Here we are, still operating on an ancient Egyptian year, Roman months, and Mesopotamian weeks; and ordinarily we get along quite nicely.

This little book reveals, I think, that Rome was our chief benefactor: not only did she bring together the various elements involved, fusing the science of the East with her own beliefs and universalizing both; but to her we owe so much in matters of linguistic and chronological detail, e.g. the names and order of the months, the distribution of days within the months, even to some extent the dating of Christian festivals like Christmas and Easter, and in Germanic guise the names of the days of the week. Moreover, examining this debt leads us into a closer awareness of our Roman legacy in other areas of life, especially religion, itself a Roman word along with sacred, profane, temple, pontiff and a host of other Latin terms which we have inherited from the people who created our calendar. But even the word money is explained in the context of such a study! And of course the source of much of our vocabulary in law and politics is evident at numerous points.

These remarks contain a pretty broad hint that the study of the Roman calendar is not a confining or narrow subject and it is not, either as an introduction to ancient mythology, legend, religion, history, and semantics, or as the basis for advanced studies in cultural anthropology. It is no accident that Sir James Frazer's *Golden Bough*, still a classic in this field, was inspired by an interest in the strange rites of Diana at Lake Nemi.

I have tried to limit this work to the bare facts and just enough interpretation—most of it merely reported—to make sense of the facts. I have injected my own views on only one subject, the origin of the calendar, because I have not encountered any other convincing explanation. I am deeply indebted to all of the books listed in the Bibliography, par-

ticularly to those by the late Warde Fowler and to those by
Professor Rose, whose learning and originality deserve great
respect. This little book is meant to be a curious and useful
handbook for the uninstructed, the partially instructed, and
even the better instructed who know, as I do, how difficult it
sometimes is to find the right information quickly and in one
packet. And that is really the secret of my composition: it was
made for my own use but I shall be pleased to have others
share it. I am grateful to my wife for the drawings which
illustrate points in the Introduction.

Tufts University

1955

Van L. Johnson

PREFACE TO SECOND EDITION

It is gratifying to know that this little book has been useful
to so many readers. I have tried to make this second edition
more instructive by touching on a few additional questions
which thoughtful persons have raised. The welcome and long-
awaited new edition of the Roman calendars by Professor
Degrassi (*Inscriptiones Italiae*, vol. 13, fasc. 2) in 1963 has
been of great assistance in supplementing and correcting
certain matters in my earlier version. I have also profited, I
think, from my own investigation of certain festivals and their
relation to the gestation cycles of domesticated animals. *Quae
vobis omnia prosint!*

Tufts University

1969

Van L. Johnson

II. INTRODUCTION

The Roman calendar is essentially a list of festivals and anniversaries on some of which it was *fas* or right to do legal business in the assembly: these "right days" or *dies fasti* gave their name to the calendar which was commonly referred to simply as *fasti*. The word calendar itself is a term of later origin derived from *Kalendae*, the name of the first day of each month. *Fas* was divine right as opposed to human right or law, viz. *ius*, the root of our word justice. *Fas* embraced all *sacra*, i.e. holy and unholy things involved in man's connections or "bonds," *religiones*, with the gods. A thing was sacred or *sacrum* because it had some relation to divinity or *numen*: this might also cause it to be profane, something to be kept *pro fano*, "in front of," i.e. "outside the shrine." The gods were *numina* or powers of assent whose cooperation was secured by the performance of certain rituals, *sacra*, either public or private. Public rites included festivals, *feriae;* games, *ludi*, either in the circus or in a theatre; and banquets, *epula*. Private rites involved domestic worship, especially of the dead and of the household gods, and ceremonies of the clan, *sacra gentilicia*.

The public rites or *sacra publica* were under the supervision of priests, *sacerdotes*, of whom the *flamines* ("fire-tenders"?) served single deities while the *pontifices* ("bridge-builders"?) might participate in various cult rites. The three principal *sacerdotes* were the *pontifex maximus*, the high pontiff of the college or *collegium* of *pontifices;* the *rex sacrorum* or king of the rites, who retained certain religious powers once belonging to the early kings; and the *flamen Dialis* or priest of Jove. The pontiffs were usually chosen by co-option of their colleagues; the *flamines* were elected in the *comitia tributa* or assembly by tribes; and the *rex sacrorum* was elected by a special assembly called the *comitia calata*, i.e. a "summoned assembly." The *pontifices* might hold civil offices as well, while the *pontifex maximus* and the *flamen Dialis* were automatically members of the senate. The *pontifices*, 4 to 16 in number at various times in history, met in the *regia* or "palace," the official residence of the *pontifex maximus* and of the *rex sacrorum*. On the Nones in each month, the *rex sacrorum* announced the festivals for that month. His wife, the *regina sacrorum* or queen of rites, sacrificed to Juno in the *regia* on the Kalends, i.e. the first of each month. The major *flamines* were those for Jupiter, Mars, and Quirinus; the lesser *flamines*, 8 in number, worshipped deities for the most part obscure or forgotten. The *flamen Dialis* and his wife, the *flaminica Dialis*, lived under severe restrictions.

The *flamen Dialis* sacrificed to Jupiter on the Ides of the month, and his wife sacrificed to Jupiter on market days or *nundinae*.

Of great importance too were the augurs, *augures*, 2 to 15 in number, who met on the Nones of each month to take the auspices or *auspicia* ("bird-watchings") : facing south, they marked out the heavens into regions with a wand or *lituus*, and interpreted the flight of birds through these *templa* or districts of the sky. Their control of portents and inaugurations gave them great political power at certain periods. The chief priestesses of Rome were the Vestal Virgins, *virgines Vestales*, 2 to 6 in number, who tended the eternal fire and the storeroom or *penus* of Vesta's temple. They were chosen at an early age by the *pontifex maximus*, and served for thirty years under his jurisdiction; they lived, under vows of chastity, in a kind of nunnery, the *atrium Vestae* or "Vesta's house." Other priesthoods and positions of priestly character were those of the *epulones* who served the banquets or *epula* of Jove; the *duumviri* (later *decemviri* and *quindecimviri*) *sacris faciundis*, i.e. the two, ten, or fifteen men to perform rites : as priests of Apollo they cared for the Sibylline books and superintended the *ludi* or games in his honor; the *fetiales*, 20 in number, who declared war, established peace, and made *foedera* or treaties; the *haruspices* or "gut-gazers," of Etruscan origin, who read portents in the *exta* or entrails of animals; the *Salii* or Leaping Priests of Mars and Quirinus; the *Fratres Arvales* or Brothers of the Ploughed Fields; and the *Luperci* or Wolf-Priests.

The rites or *sacra* supervised by these various priesthoods, along with certain political accretions, constitute the Roman calendar as preserved in ancient sources. It is therefore fundamentally a religious calendar, but economic and astronomical factors were also involved in its origin and development as a mode of time-reckoning. The Roman calendar was not published until 304 B.C. and it is known to us only through evidence of much later date; but its earlier history can be reconstructed from the facts available.

For most primitive peoples the sun measures only the day; the moon, with its discernible phases, is the first measurement of periods beyond that. In addition, they often have market-days with regular intervals between them. The Romans follow this pattern; and it is possible that the oldest Roman calendar was simply an attempt to mesh a 30-day lunation with an 8-day market week. The 8-day week may have been determined by the interval required for the processing of goat's cheese. The market-days were called *nundinae* or "ninth-days," i.e. the eighth-days by our method of reckoning which is not inclusive like the Roman. The meshing of these two time units could be soonest accomplished in 4 months of 30 days each, i.e. a year or cycle, *annus*, of 120 days. To these four original months, ultimately named March, April, May, and June, 6 months, simply numbered from 5 to 10—Quintilis, Sextilis, September, October, November, and December—were added, traditionally by Romulus. There is some reason for thinking that March was originally called Caprotinus and June, Fabarius. These names, like April and May, possibly refer to the breeding and raising of pigs, since the 120-day year would correspond to the gestation cycle of the pig.

The new year of 10 lunations, corresponding roughly to the gestation period in cattle and in human beings, was augmented by 4 days to give a multiple of 8 for the total number of days in a year, viz. 304, so that the 8-day week would still mesh with months. These 4 extra days were added, one each, to the months of March, May, July, and October which continued throughout Roman history, to have their Ides reckoned on the basis of a 31-day month. The 31-day months are not exact alternates, because—on one view—a September of 31 days would have produced a nundinal Nones in that month; this would have caused confusion because both terms, *nundinae* and *Nonae*, have the same meaning—"ninth-days"—but serve different functions in the calendar. According to one view, this 10-month calendar simply left the unproductive days of winter uncounted; another interpretation makes this year operate continuously without regard to the sun's course.

Numa, the second king of Rome, is credited with instituting the first true lunar year by adding 50 days to the calendar of Romulus. To equalize the distribution of 354 days over 12 months, he subtracted one day from each of the 30-day months, added these to his 50 new days, and divided the sum, 56, into two new months of 28 days each, viz. January and February. Then, for some superstitious reason, variously explained, he added another day to January, creating a year of 355 days which continued until the time of Caesar. But neither 354 nor 355 is a multiple of 8, so "Numa's" year could mesh with the Roman 8-day week only by introducing or intercalating extra days toward the end of December, if the year started with January, toward the end of February, if the year started with March: the Saturnalia of December 17 and the Terminalia of February 23 were probably the starting-points of such intercalary periods. It appears that "Numa" meant his calendar to begin with January, but tradition retained March as the first month, as late as 153 B.C. for some purposes.

During the Republic, intercalation came to be used for a different purpose, viz. to bring the lunar year into accord with the solar year of 365 or 366 days. An intercalary month of 22 or 23 days, called Mercedonius, was added in alternate years after February 23, the day of the Terminalia, and the last 5 days of February were absorbed as the last 5 of Mercedonius. This device was so often neglected, however, or corrupted by priestly or political abuses that the calendar had very little relation to sun's course when Julius Caesar and his learned adviser, Sosigenes, introduced those reforms which are still the basis of our calendar. Caesar first extended the year 46 B.C. to 445 days, thus bringing the old calendar into agreement with astronomical observations; then, on January 1, 45 B.C. he introduced a solar calendar of 365¼ days: the fourth parts were allowed to accumulate and produce a year of 366 days once every four years. To achieve the 10 new days of a normal year (365 minus 355), Caesar added 2 days each to January, August, and December; one day each to April, June, September, and November; and the extra day for leap years was inserted after February 23 as February 24 repeated, i.e. *bissextilis,* the "twice sixth-day" before the first of March: hence we speak of a bissextile year. Caesar's additions did not alter the Nones or Ides of any month or affect the position of festivals: those subsequent to the Ides of a month remained at the same interval from the Ides, even though this involved giving them a new date in relation to the next Kalends.

Caesar's calendar was 11 minutes, 14 seconds too long, so Pope Gregory made a slight adjustment in 1582 A.D., dropping 10 excessive days at once and stipulating that leap year be observed in centesimal years only when they are divisible by 400. These corrections were not accepted in Great Britain and the American colonies until 1752 A.D.

Roman years were reckoned by their number "from the founding of the city," *ab urbe condita,* an event which the Romans dated in the year 753 B.C.: therefore, to transfer dates a.u.c. to our own system of dating, we must subtract a.u.c. dates below 754 from 754 to give dates B.C., or subtract 753 from a.u.c. dates exceeding 753 to give dates A.D. During the period of the Roman Republic, years were also identified by consulships, which were annual magistracies and thus unambiguous for this purpose. During the Empire, when the consulship became only an honorary office and irregular in tenure, years were identified by various titles and powers conferred upon the Emperor and renewed annually.

Of the Roman months, January, March, and June were probably named for gods or goddesses—Janus, Mars, and Juno respectively (but see the Glossary for references to notes concerning these deities); February was named for *februa* or "purifications" which were common in the ceremonies of that month; April and May, according to recent interpretation, were months of the "boar," *aper* and of the "sow," *maia;* the last six months were simply numbered from March as the first month, but Quintilis and Sextilis, the Fifth and Sixth, were renamed July and August in honor of Julius Caesar and Augustus in 44 B.C. and 8 B.C. respectively.

Within a month, the Romans of the historical period calculated dates in reference to days called the Kalends, the Nones, and the Ides: the Kalends was always the first day of the month, giving us our word calendar; the Nones and the Ides fell on the fifth and thirteenth days respectively in all months except March, May, July, and October when they occurred on the seventh and fifteenth. These three days in the month were named, and all other days were computed retrogressively from them: thus any day after the Ides of March is so many days before the Kalends of April. By Roman calculation, which reckoned both ends of an arithmetical series, this is always one more day than we should include. Thus March 13, for example, is expressed as 3 days before the Ides (March 15) of March; and this would be most commonly stated in Latin by the odd phrase, *ante diem tertium Idus Martias*, where the month name is an adjective limiting *Idus*, and the same preposition, *ante*, governs two accusatives. This is odd Latin, but perhaps we should regard the whole matter as an early grammatical pattern. The same might be said for a phrase expressing March 14, which would be described as *pridie Idus Martias*, "the day before the Ides of March," where we have an adverb followed by the accusative. The adverb, of course, is derived from a noun in the ablative case; and the ablative case is always used for named dates, e.g. *Idibus Martiis* for the Ides or 15th of March. Abbreviated forms of all dates will be found on pages 20-31.

The Ides were originally the "Dividers," dividing in half the 30-day month; they always occur 16 days before the last day of the month, because half of 30 is 16, not 15, on the duodecimal system which the Romans used in computing fractions. The Ides became prominent perhaps because they were sometimes concurrent with the days of the full moon; they were sacred to Jupiter. The Nones were the "ninth-days" (i.e. the eighth-days) before the Ides, marking sometimes the days of the moon's first quarter; and the Kalends or "callings" of the Nones as fifth or seventh day of the month were sometimes the days of the visible new moon: the "calling" was made by a pontiff who addressed Juno Covella either five or seven times to date the Nones as fifth or seventh. Thus all Kalends were sacred to Juno. On the Nones the *rex sacrorum* announced the festivals or *feriae* for the month; this explains why almost all *sacra publica* occur after the Nones of a month.

The days of the Roman 8-day week were simply lettered from A to H. It has been variously maintained that A, H, or F was the original nundinal letter, i.e. the letter which named the *nundinae* or market-days; but the letter remained constant only as long as the number of days in a year was divisible by 8, or while intercalary devices were used to accomplish the same end by adding days to the week as well as to the year. In the later calendars it appears that the nundinal letter shifted from year to year, since the first day of each year was always marked A and the total number of days in a year was not a multiple of 8. After establishment of the Ecclesiastical calendar in 325 A.D., the Christian Church borrowed this practice of a shifting week-day letter: the "dominical" or "Lord's day" letter is still used for determining the date of Easter.

The 7-day planetary week was not common in Rome or the West until the third century A.D., though it appears to have existed alongside the Roman 8-day week in a Sabine calendar of the first century A.D. The planetary week had its origins in the Eastern Mediterranean where Babylonian astrology, the Hebrew Sabbath-week, and Egyptian astronomy combined to formulate and confirm it. The nature of the planets was first discovered in Mesopotamia where an intense interest in the heavens gave rise to the pseudo-science of astrology: the sun and moon, however, were included in a list of 7 "planets" or "travellers," and the earth was omitted as being the stationary center of a geocentric universe. Saturn, Jupiter, Mars, Venus, and Mercury (to use the Latin names still current) were properly located, at least in respect of their relative distances from the earth. Pluto, Neptune, and Uranus were, of course, unknown, so the seven "planets" were Saturn (outermost), Jupiter, Mars, Sol, Venus, Mercury, and Luna in that

---

ANCIENT "PLANETS"

1	2	3	4	5	6	7
Saturn	Jupiter	Mars	Sun	Venus	Mercury	Moon

PLANETS KNOWN TODAY

1	2	3	4	5	6	7	8	9
Pluto	Neptune	Uranus	Saturn	Jupiter	Mars	Earth	Venus	Mercury

---

order. The planets were conceived as moving in spheres or orbits around the earth and passing through twelve constellations or "fixed stars" which made a zodiac or "belt of animals" around the heavens with the sun's path or ecliptic as the middle line. Since the sun travels through all twelve signs of the zodiac in one year and remains about the same length of time in each, this system evolved something like a solar month which could be related to the equinoxes and the solstices. (See TABLE I, page 13: the dates for the sun's entrances are generalized.)

PLANETS and ZODIAC

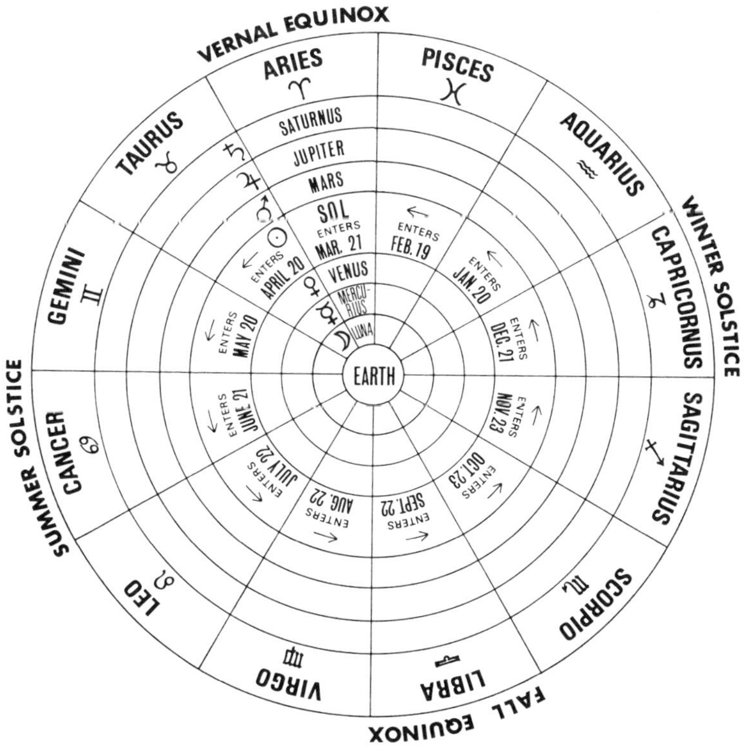

TABLE I

Reprinted from *Archaeology*, Vol. 21, No. 1 (January 1968) page 19.

PERMUTATION CYCLE
PLANETS and 24-HOUR DAY

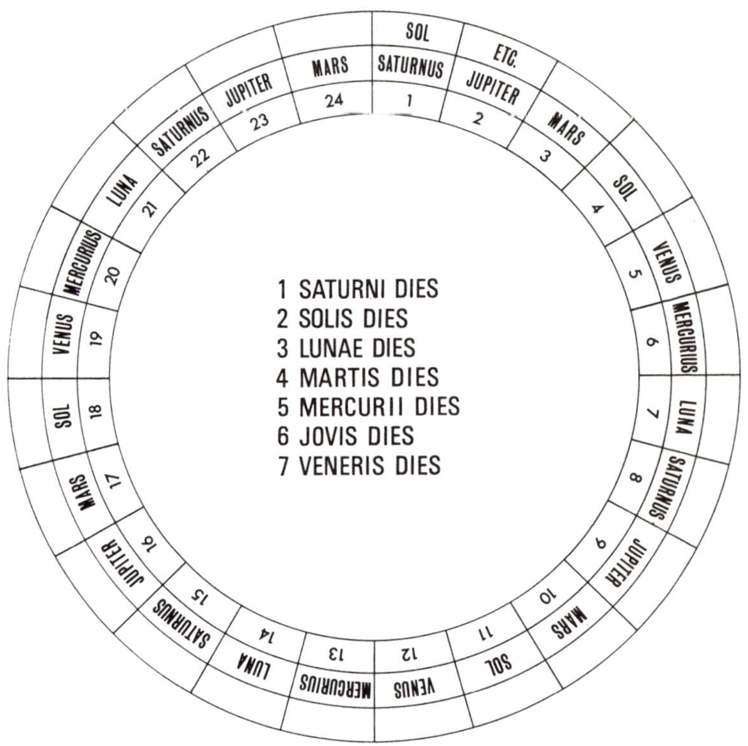

TABLE II

The innermost circle represents the 24 hours of the day; the second circle, the names of these hours on the first day of the week; the third circle, the names of the hours on the second day of the week, where the naming continues as in the second circle. Plotting additional circles for the third to the seventh days would lead to the arrangement of first hours indicated within the circle for all seven days of the week. A circle for the eighth day would bring us back to Saturn for the first hour, and thus to the beginning of a new week.

Reprinted from *Archaeology*, Vol. 21, No. 1 (January 1968) page 20.

The zodiac is so called because it consists of constellations which suggested various "living things" by the configurations of stars within them. These constellations or "fixed stars" still have Latin names because the "animals" involved were associated with Greek myths familiar to the Romans. *Aries* is the Ram whose golden fleece was sought by Jason. *Taurus* is the Bull whose shape Zeus assumed when he seduced Europa. *Gemini* are the Twins, Castor and Pollux, brothers of Helen, whom Zeus put in the heavens when the immortal Pollux chose not to outlive his mortal brother, Castor. *Cancer* is the Crab sent by Hera to bite Hercules when he was fighting the Lernaean hydra. *Leo* is the Nemean Lion slain by Hercules. *Virgo* is Astraea, the goddess of Justice, last of the gods to leave the earth when Jupiter expelled Saturn and the Golden Age came to an end. *Libra* is the Balance or Scales of Justice belonging to Virgo. *Scorpio* is the Scorpion sent by Artemis to bite Orion when he assaulted her. *Sagittarius* or the Archer is Chiron the centaur, teacher of Hercules, Achilles, and other heroes. *Capricornus* is the Goat, Amalthea, whose milk sustained the infant Zeus on the island of Crete. *Aquarius* or the Water-Bearer is Ganymedes, the Trojan prince beloved by Zeus and carried to heaven to become the cup-bearer of the gods. *Pisces* are the Fish into which Aphrodite and Eros were changed when they fled to Egypt to escape the monster, Typhon.

But in Egypt the planetary system underwent further development, as the 7 "planets" were meshed with a 24-hour day: assigning Saturn to the first hour of the first day, Jupiter to the second hour, Mars to the third, etc., introduces Sol at the first hour of the second day, Luna at the first hour of the third day, and so on until each "planet" has been associated with that day of which it marks the first hour. Thus the 7 "planets" name the 7 days of the week in this order: Saturn, Sol, Luna, Mars, Mercury, Jupiter, and Venus. (See TABLE II, page 14, based upon Dio Cassius, 37. 18-19.) In this sequence these 7 celestial bodies still name, with minor exceptions, the 7 days of our week; French preserves the Latin names, while German and English translate them into the names of counterparts among Germanic deities:

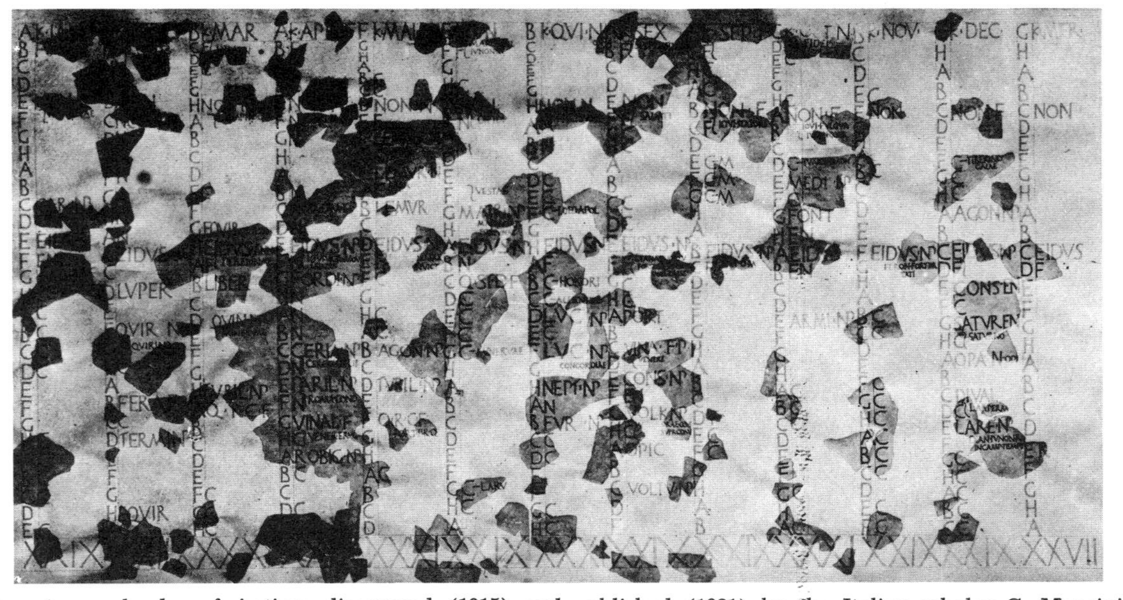

The fragmentary calendar of Antium discovered (1915) and published (1921) by the Italian scholar G. Mancini. This inscription records the oldest extant Roman calendar so far unearthed. It is beautifully lettered by the stone cutter and the letters filled in with red and black. The thirteen months (each represented by a column with the abbreviations for the name of the month at the top) include the intercalary month. The first column of letters under each month indicates the eight days of the Roman week lettered from A to H. The second column of letters abbreviates the legal status of certain days (e.g. *N* stands for *nefastus dies*) and the abbreviated words indicate the dates of festivals or dedications (e.g. *PARIL* stands for *Parilia* on April 21). Photograph courtesy of Fototeca di architettura e topografia dell' Italia antica.

Reprinted from *Archaeology*, Vol. 21, No. 1 (January 1968) page 15.

LATIN	FRENCH	GERMAN	ENGLISH
Solis dies	*dimanche	Sonntag	Sunday
Lunae dies	lundi	Montag	Monday
Martis dies	mardi	‡Dienstag	‡Tuesday
Mercurii dies	mercredi	§Mittwoch	‡Wednesday
Jovis dies	jeudi	‡Donnerstag	‡Thursday
Veneris dies	vendredi	‡Freitag	‡Friday
Saturni dies	†samedi	†Samstag	Saturday

dimanche from *dies dominica*, the Lord's Day. Sunday received this significance by an edict of Constantine in 321 A.D. and by the decrees of the Council of Nicaea (325 A.D.) which established a Sunday date for Easter.

‡The Germanic peoples identified their Tiu with Mars, Woden with Mercury, Thor-Donar with Jove, and Freia with Venus.

§*Mittwoch*, i.e. Mid-week

†*samedi* and *Samstag* from *Sabbati dies*, day of the Hebrew Sabbath.

Our 24-hour day is something of an accident: as noted above, it existed in Egypt and it seems to go back to the Egyptian "diagonal" calendars (see **TABLE III**, page 18) which date anywhere from 1800 to 1200 B.C. The Egyptian year had 12 months of 30 days each, plus 5 "epagomenal" days. The 360 days of the 12 months were divided into 36 "decades" of 10 days each, and to each decade was assigned a "decan" or constellation of which the heliacal or dawn rising marked the last hour of the night in that decade. If day and night were always of equal length, 18 decans could have been seen every night; but due to variations in the length of night and twilight, this was not always possible; and indeed only 12 decans were visible in Egypt during the short nights of summer when the dawn rising of Sirius marked the inundation of the Nile—an important matter in the ancient Egyptian year. The night was therefore divided into 12 parts or hours, and Sirius, for example, marks the 12th or last hour in the first decade, the 11th hour in the second decade, etc. In the second decade, another decan marks the 12th hour; in the next decade the 11th hour, etc. To these 12 hours of night, there were added 10 hours of daylight and 2 hours of morning and evening twilight, thus giving a total of 24 hours to each day in the usual sense of day plus night.

Our division of the hour into 60 minutes is a result of Hellenistic computations worked out on the sexagesimal system first devised by the Babylonians about 1800 to 1600 B.C.

Days in the Roman calendar were marked not only with the letters A to H, indicating the day of the market week, but also with letters denoting the juridical character of each day. These letters and their probable meanings are shown in the following list:

F for *fastus dies*, a day on which it is right to do legal business in the *comitia* or assembly, and in the praetor's court.

N for *nefastus dies*, a day on which it is wrong to transact such business.

NP for *nefastus parte*, a day of which part is wrong for the transaction of legal business.

FP for *fastus principio*, a day at the beginning of which it is right to transact legal business.

EN for *endotercisus*, i.e. *intercisus dies*, a day of three parts when it is wrong to do legal business early or late in the day, but right in the middle of the day.

C for *comitialis dies*, a day when the *comitia* or assembly will meet.

EGYPTIAN "DIAGONAL" CALENDAR

DECANS

	XXXVI			**Hours of the Night**	
	•				
	•				
	•				
	III			12	etc.
	II		12	11	etc.
Sirius	I	12	11	10	etc.
	I	II	III.................XXXVI		

DECADES

TABLE III

III. FASTI ROMANI

LEGEND

*†Events first marked with either of these symbols are repeated on days continuously marked thus.

( ) Parentheses indicate the alternate Roman dating for February in leap years.

Festivals recorded in capital letters are the so-called *Fasti Antiquissimi*, the oldest festivals still observed in later times.

Festivals are named in the Latin nominative case; deities, in the Latin dative case.

This calendar includes most important entries from calendars of Imperial date, and selected entries from the *Fasti* of Antium, the only extant calendar of Republican date.

JANUARIUS

1	Kalendis	Aesculapio Vediovi
2	IV Nonas	
3	III Nonas	*Compitalia Cicero born, 106 B.C.
4	Pridie Nonas	*
5	Nonis	*Vicae Potae
6	VIII Idus	
7	VII Idus	Iano patri circenses
8	VI Idus	
9	V Idus	AGONALIA
10	IV Idus	
11	III Idus	CARMENTALIA Iuturnalia Augustus closed the temple of Janus, 29 B.C.
12	Pridie Idus	
13	Idibus	Augustus "restored" the Republic, 27 B.C. Iovi statori circenses
14	XIX Kalendas	Mark Antony born, 82 B.C. Nero Claudius Drusus the Elder born, 38 B.C.
15	XVIII Kalendas	CARMENTALIA
16	XVII Kalendas	Octavian named Augustus, 27 B.C. Concordiae Augustae
17	XVI Kalendas	
18	XV Kalendas	
19	XIV Kalendas	
20	XIII Kalendas	
21	XII Kalendas	
22	XI Kalendas	
23	X Kalendas	
24	IX Kalendas	†Sementivae Hadrian born, 76 A.D.
25	VIII Kalendas	†
26	VII Kalendas	†
27	VI Kalendas	Castori et Polluci
28	V Kalendas	
29	IV Kalendas	
30	III Kalendas	Dedication of the *ara pacis*, 1 B.C. Livia born
31	Pridie Kalendas	Antonia Minor born

FEBRUARIUS

1	Kalendis	Iunoni Sospitae
2	IV Nonas	
3	III Nonas	
4	Pridie Nonas	
5	Nonis	Concordiae in arce Augustus named *pater patriae*, 2 B.C.
6	VIII Idus	
7	VII Idus	
8	VI Idus	
9	V Idus	
10	IV Idus	
11	III Idus	
12	Pridie Idus	
13	Idibus	*Parentalia Fauno in insula
14	XVI Kalendas	*
15	XV Kalendas	*LUPERCALIA
16	XIV Kalendas	*
17	XIII Kalendas	*QUIRINALIA Fornacalia Stultorum feriae
18	XII Kalendas	*
19	XI Kalendas	*
20	X Kalendas	*
21	IX Kalendas	*FERALIA
22	VIII Kalendas	Caristia
23	VII Kalendas	TERMINALIA
24	VI Kalendas	REGIFUGIUM
25	V (bis VI) Kalendas	
26	IV (V) Kalendas	
27	III (IV) Kalendas	EQUIRRIA Constantine born, 274 A.D.
28	Pridie (III) Kalendas	
29	(Pridie) Kalendas	

MARTIUS

1	Kalendis	Feriae Marti Iunoni Lucinae Matronalia
2	VI Nonas	
3	V Nonas	
4	IV Nonas	
5	III Nonas	
6	Pridie Nonas	Augustus made *pontifex maximus*, 12 B.C.
7	Nonis	Vediovi
8	VIII Idus	
9	VII Idus	Arma ancilia movent
10	VI Idus	Tiberius made *pontifex maximus*, 15 A.D.
11	V Idus	
12	IV Idus	
13	III Idus	
14	Pridie Idus	EQUIRRIA Feriae Marti Mamuralia
15	Idibus	Annae Perennae Caesar assassinated, 44 B.C.
16	XVII Kalendas	*Sacra Argeorum
17	XVI Kalendas	*LIBERALIA AGONALIA
18	XV Kalendas	
19	XIV Kalendas	†QUINQUATRUS Feriae Marti
20	XIII Kalendas	†
21	XII Kalendas	†
22	XI Kalendas	†
23	X Kalendas	†TUBILUSTRIUM
24	IX Kalendas	Q.R.C.F.
25	VIII Kalendas	Hilaria
26	VII Kalendas	
27	VI Kalendas	Lavatio Caesar recovered Alexandria, 47 B.C.
28	V Kalendas	
29	IV Kalendas	
30	III Kalendas	Iano, Concordiae, Saluti, et Paci
31	Pridie Kalendas	Lunae in Aventino

APRILIS

1	Kalendis	Veneralia
		Fortunae virili in balneis
2	IV Nonas	
3	III Nonas	
4	Pridie Nonas	Magnae Matri
		*Ludi Megalesiaci sive Megalesia
5	Nonis	*Fortunae publicae citeriore in colle
6	VIII Idus	*Battle of Thapsus, 46 B.C.
7	VII Idus	*
8	VI Idus	*
9	V Idus	*
10	IV Idus	*Magnae Matri
11	III Idus	Septimius Severus born, 146 A.D.
12	Pridie Idus	†Ludi Cereales
13	Idibus	†Iovi Victori
14	XVIII Kalendas	†
15	XVII Kalendas	†FORDICIDIA
16	XVI Kalendas	†*Dies imperii* of Augustus, 29 B.C.
17	XV Kalendas	†
18	XIV Kalendas	†
19	XIII Kalendas	†CEREALIA
		Cereri, Libero, et Liberae
		Ludi circenses
20	XII Kalendas	
21	XI Kalendas	PARILIA
22	X Kalendas	
23	IX Kalendas	VINALIA
24	VIII Kalendas	Feriae Latinae
25	VII Kalendas	ROBIGALIA
26	VI Kalendas	Marcus Aurelius born, 121 A.D.
27	V Kalendas	
28	IV Kalendas	*Ludi Florae sive Floralia
29	III Kalendas	*
30	Pridie Kalendas	*

MAIUS

1	Kalendis	*Ludi Florae Bonae Deae Laribus praestitibus
2	VI Nonas	*
3	V Nonas	*
4	IV Nonas	
5	III Nonas	
6	Pridie Nonas	
7	Nonis	†Vestal virgins prepare the *mola salsa*
8	VIII Idus	†
9	VII Idus	†LEMURIA
10	VI Idus	†
11	V Idus	†LEMURIA
12	IV Idus	†Ludi Marti in circo Marti Ultori
13	III Idus	†LEMURIA
14	Pridie Idus	†Marti invicto
15	Idibus	Sacra Argeorum Feriae Iovi, Mercurio, Maiae
16	XVII Kalendas	
17	XVI Kalendas	
18	XV Kalendas	
19	XIV Kalendas	
20	XIII Kalendas	
21	XII Kalendas	AGONALIA Vediovi
22	XI Kalendas	
23	X Kalendas	TUBILUSTRIUM Volcano
24	IX Kalendas	Q.R.C.F. Germanicus born, 15 B.C.
25	VIII Kalendas	Fortunae publicae populi Romani
26	VII Kalendas	
27	VI Kalendas	
28	V Kalendas	
29	IV Kalendas	Ambarvalia
30	III Kalendas	
31	Pridie Kalendas	

JUNIUS

1	Kalendis	Iunoni Monetae
		Carnae
		Kalendae fabariae
		Tempestati
		Marti ad portam Capenam
2	IV Nonas	
3	III Nonas	Bellonae in circo Flaminio
4	Pridie Nonas	Herculi magno custodi
5	Nonis	Dio Fidio in colle
6	VIII Idus	
7	VII Idus	Vesta aperitur
		Ludi piscatorii
8	VI Idus	Menti in Capitolio
9	V Idus	VESTALIA
10	IV Idus	
11	III Idus	MATRALIA
		Fortunae
		Concordiae
12	Pridie Idus	
13	Idibus	Feriae Iovi
		Quinquatrus minusculae
14	XVIII Kalendas	
15	XVII Kalendas	Q. ST. D. F.
16	XVI Kalendas	
17	XV Kalendas	
18	XIV Kalendas	Annae sacrum
19	XIII Kalendas	Minervae in Aventino
20	XII Kalendas	Summano ad circum maximum
21	XI Kalendas	
22	X Kalendas	
23	IX Kalendas	*Dies ater*
24	VIII Kalendas	Forti Fortunae trans Tiberim
25	VII Kalendas	
26	VI Kalendas	
27	V Kalendas	Iovi statori
		Laribus publicis in summa sacra via
28	IV Kalendas	
29	III Kalendas	Quirino in colle
30	Pridie Kalendas	Herculi Musarum

QUINTILIS (JULIUS)

1	Kalendis	Felicitati in Capitolio
2	VI Nonas	
3	V Nonas	
4	IV Nonas	Ara pacis Augustae constituta, 13 B.C.
5	III Nonas	POPLIFUGIA Feriae Iovi
6	Pridie Nonas	*Ludi Apollinares Fortunae Muliebri
7	Nonis	*Nonae Caprotinae Ancillarum feriae Sacrificium Conso Palibus II
8	VIII Idus	*Vitulatio
9	VII Idus	*
10	VI Idus	*
11	V Idus	*
12	IV Idus	*Julius Caesar born, 102 (?) B.C.
13	III Idus	*
14	Pridie Idus	†Mercatus
15	Idibus	†Equitum probatio
16	XVII Kalendas	†
17	XVI Kalendas	†
18	XV Kalendas	†*Dies ater*
19	XIV Kalendas	†LUCARIA
20	XIII Kalendas	*Ludi victoriae Caesaris
21	XII Kalendas	*LUCARIA
22	XI Kalendas	*Concordiae
23	X Kalendas	*NEPTUNALIA
24	IX Kalendas	*
25	VIII Kalendas	*FURRINALIA
26	VII Kalendas	*
27	VI Kalendas	*
28	V Kalendas	*
29	IV Kalendas	*
30	III Kalendas	*Fortunae huiusque diei in campo
31	Pridie Kalendas	

SEXTILIS (AUGUSTUS)

1	Kalendis	Spei ad forum holitorium
		Marti Ultori
		Claudius born, 10 B.C.
2	IV Nonas	
3	III Nonas	
4	Pridie Nonas	
5	Nonis	Saluti in colle Quirinali
6	VIII Idus	
7	VII Idus	
8	VI Idus	
9	V Idus	Soli Indigiti in colle Quirinali
		Battle of Pharsalus, 48 B.C.
10	IV Idus	
11	III Idus	
12	Pridie Idus	Herculi invicto ad circum maximum
13	Idibus	Feriae Iovi
		Dianae in Aventino
		Vortumno in Aventino
		Herculi invicto ad portam Trigeminam
		Castori Polluci in circo Flaminio
		Fiorae ad circum maximum
14	XIX Kalendas	
15	XVIII Kalendas	
16	XVII Kalendas	
17	XVI Kalendas	PORTUNALIA
		Tiberinalia
		Iano ad theatrum Marcelli
18	XV Kalendas	Divo Iulio
19	XIV Kalendas	VINALIA
		Feriae Iovi
		Veneri ad circum maximum
		First consulship of Augustus, 43 B.C.; died, 14 A.D.
20	XIII Kalendas	
21	XII Kalendas	CONSUALIA
22	XI Kalendas	
23	X Kalendas	VOLCANALIA
24	IX Kalendas	Mundus patet
25	VIII Kalendas	OPICONSIVIA
26	VII Kalendas	
27	VI Kalendas	VOLTURNALIA
28	V Kalendas	
29	IV Kalendas	
30	III Kalendas	
31	Pridie Kalendas	Caligula born, 12 A.D.

SEPTEMBER

1	Kalendis	Iovi Tonanti in Capitolio Iovi Libero, Iunoni Reginae in Aventino
2	IV Nonas	Battle of Actium, 31 B.C.
3	III Nonas	
4	Pridie Nonas	*Ludi Romani
5	Nonis	*Iovi Statori
6	VIII Idus	*
7	VII Idus	*
8	VI Idus	*
9	V Idus	*Aurelian born, 214 A.D.
10	IV Idus	*
11	III Idus	*
12	Pridie Idus	*
13	Idibus	*Iovi Epulum
14	XVIII Kalendas	*Equorum Probatio
15	XVII Kalendas	*
16	XVI Kalendas	*
17	XV Kalendas	*
18	XIV Kalendas	*Trajan born, 53 A.D.
19	XIII Kalendas	*Antoninus Pius born, 86 A.D.
20	XII Kalendas	†Mercatus
21	XI Kalendas	†
22	X Kalendas	†
23	IX Kalendas	†Marti, Neptuno in Campo Apollini in theatro Marcelli Iovi Statori Iunoni Reginae ad circum Flaminium Augustus born, 63 B.C.
24	VIII Kalendas	
25	VII Kalendas	
26	VI Kalendas	Veneri Genetrici in foro Caesaris
27	V Kalendas	
28	IV Kalendas	
29	III Kalendas	
30	Pridie Kalendas	

OCTOBER

1	Kalendis	Tigillo sororio ad compitum Acili Fidei in Capitolio Alexander Severus born, 208 A.D.
2	VI Nonas	
3	V Nonas	*Ludi Augustales
4	IV Nonas	*Ieiunium Cereris
5	III Nonas	*Mundus patet
6	Pridie Nonas	*Dies ater
7	Nonis	*Iovi Fulguri Iunoni Curriti in Campo
8	VIII Idus	*
9	VII Idus	*Apollini in Palatio
10	VI Idus	*Iunoni Monetae
11	V Idus	*MEDITRINALIA
12	IV Idus	*Augustalia
13	III Idus	FONTINALIA
14	Pridie Idus	
15	Idibus	Feriae Iovi October Horse Vergil born, 70 B.C.
16	XVII Kalendas	
17	XVI Kalendas	
18	XV Kalendas	
19	XIV Kalendas	ARMILUSTRIUM
20	XIII Kalendas	
21	XII Kalendas	
22	XI Kalendas	
23	X Kalendas	
24	IX Kalendas	
25	VIII Kalendas	†Ludi Victoriae Sullanae
26	VII Kalendas	†
27	VI Kalendas	†
28	V Kalendas	*†Isia
29	IV Kalendas	*†
30	III Kalendas	*†
31	Pridie Kalendas	*†

NOVEMBER

1	Kalendis	*†(See Oct. 25 and 28)
2	IV Nonas	
3	III Nonas	
4	Pridie Nonas	*Ludi Plebeii
5	Nonis	*
6	VIII Idus	*Agrippina the Younger born, 16 A.D.
7	VII Idus	*
8	VI Idus	*Mundus patet. Nerva born, 32 A.D.
9	V Idus	*
10	IV Idus	*
11	III Idus	*
12	Pridie Idus	*
13	Idibus	*Feroniae in Campo Fortunae primigeniae in colle Feriae Iovi: Iovi Epulum
14	XVIII Kalendas	*Equorum probatio
15	XVII Kalendas	*
16	XVI Kalendas	*Tiberius born, 42 B.C.
17	XV Kalendas	*Vespasian born, 9 A.D.
18	XIV Kalendas	†Mercatus
19	XIII Kalendas	†
20	XII Kalendas	†
21	XI Kalendas	
22	X Kalendas	
23	IX Kalendas	
24	VIII Kalendas	
25	VII Kalendas	
26	VI Kalendas	
27	V Kalendas	
28	IV Kalendas	
29	III Kalendas	
30	Pridie Kalendas	

DECEMBER

1	Kalendis	Neptuno Pietati Fortunae Muliebri
2	IV Nonas	
3	III Nonas	Sacra Bonae Deae
4	Pridie Nonas	
5	Nonis	Faunalia rustica
6	VIII Idus	
7	VII Idus	
8	VI Idus	Tiberino in insula
9	V Idus	
10	IV Idus	
11	III Idus	AGONALIA Septimontium
12	Pridie Idus	Conso in Aventino
13	Idibus	Telluri et Cereri in Carinis
14	XIX Kalendas	
15	XVIII Kalendas	CONSUALIA Nero born, 37 A.D.
16	XVII Kalendas	
17	XVI Kalendas	*SATURNALIA
18	XV Kalendas	*
19	XIV Kalendas	*OPALIA
20	XIII Kalendas	*
21	XII Kalendas	*DIVALIA
22	XI Kalendas	*Laribus permarinis
23	X Kalendas	*LARENTALIA Feriae Iovi
24	IX Kalendas	
25	VIII Kalendas	Natalis invicti solis
26	VII Kalendas	
27	VI Kalendas	
28	V Kalendas	
29	IV Kalendas	
30	III Kalendas	Titus born, 39 A.D.
31	Pridie Kalendas	

IV. NOTES

arranged by months and days

JANUARY

1. *Aesculapio:* rites for Aesculapius or Asclepius, the Greek god of medicine, whose temple on the Tiber Island was established in 291 B.C. The cult was introduced from Epidaurus to avert a plague in Rome. The Tiber Island still retains its connection with the healing arts, since the hospital of S. Bartolomeo is located there.

   Vediovi: rites for Vediovis, Veiovis, or Vedius, an ancient Italian god usually identified with Apollo or the youthful Jupiter. Like Aesculapius and Faunus, he was worshipped on the Tiber Island. He also had a temple on the Capitoline hill. He was represented as a young archer with a goat, and a goat was the victim of sacrifice *ritu humano.* Ovid explains the name Vediovis as meaning "not-Jove" or "Jove in his bad aspect." He has also been interpreted as an "anti-Jove," i.e. a god of the underworld as opposed to the sky-god, Jupiter; in this capacity, Vediovis has been identified with Soranus, the god of Mt. Soracte, and paired with the mysterious goddess, Feronia (cf. November 13). A recent view interprets his name as "yearling-god" and associates him particularly with the beginning of years and with gestation cycles. Vediovis was worshipped as Father Vediovis by the Julian family at Bovillae. Cf. March 7, May 21, and July 7.

3. *Compitalia:* rites at the *compita* or cross-roads where shrines of the Lares Compitales were erected. The origin of the Compitalia was ascribed to Rome's Etruscan kings, Servius Tullius or Tarquinius Superbus. Originally broken ox-yokes were consecrated by farmers at the *compita.* In addition, puppets were hung at the cross-roads, one for each free person in a family; and a ball of wool was suspended for each slave. These latter customs have been interpreted as offerings to the Lares in the role of sinister spirits which haunted the cross-roads. It has been suggested that the Lares were originally deities of the farmland introduced to the household by farm-hands. The *Lares familiares* or household gods, whatever their origin, were kindly spirits; for this reason, the word Lares has even been connected with *lascivi,* meaning the "jolly ones." Some scholars trace the origin of the household Lares to early household burials. The mother of the Lares is variously described as Lara, Mania ("the good one"?), a goddess of death, or Acca Larentia (cf. December 23).

Cicero was born on this day in 106 B.C. at Arpinum which was also the birthplace of Marius, the great popular leader. His life was too full and his works too numerous to summarize here: lawyer, statesman, philosopher, and humanist, justly famous for his erudition and his rhetoric, he played a major role in the defense of Republican ideals and in the transmission of Greek ideas. Vanity and compromise often weakened his political position, but he was unflinching in his famous attacks on Verres, Catiline, and Antony. For the last, he paid with his life in 43 B.C. when he was executed by Antony's soldiers.

5. *Vicae Potae:* rites for Vica Pota, explained in antiquity as a goddess of Victory and Possession, as if from *vincere* and *potiri,* or as a goddess of Food and Drink, as if from *victus* and *potus.* She was represented as the mother of Diespiter, an old name for Jupiter; and she had a shrine on the Velian hill. These rites of January 5 are listed only in the Republican calendar of Antium; apparently they ceased to be observed in Imperial times.

7. *Iano patri circenses:* circus games in honor of Father Janus (cf. January 11). A consul threw out a *mappa* or napkin to start the races—a tradition traced back to the Tarquins (cf. January 13 and April 19).

9. *AGONALIA:* a ram was sacrificed by the *rex sacrorum,* perhaps to Janus, Inuus, or Indiges, but we know nothing more about this ancient rite. Even the meaning of the word *agonia* or *agonalia* is lost; for Ovid's derivations, see the *Fasti* 1. 318-338. A recent interpretation of *agonium* relates it to *annus* as a year-cycle for the breeding of various animals, and suggests that it was held for one or more deities of this cycle, viz. *indigetes* or yearling gods. These rites of January would have fallen in May in a 4-month year, and they may have constituted both the beginning and end of breeding cycles for pigs.

11. *CARMENTALIA:* the rites of Carmenta or Carmentis, traditionally the mother of Evander who was thought to have founded a Greek city on the site of Rome long before Rome itself was established. In antiquity Carmenta's name was connected with *carmen,* "chant," and she was accorded the gift of prophecy; her name, in the plural, was identified with the obscure deities, Porrima or Prorsa, and Postverta who were probably goddesses of child-birth. The name Carmentis appears once as a masculine form, but this is inexplicable. Only women were admitted to her temple at the Porta Carmentalis, and nothing made of leather could

be brought into her shrine. She has also been interpreted as a nymph or water-sprite. There was a *flamen Carmentalis*, indicating the great antiquity of these rites.

Iuturnalia: rites for Juturna, the deity of a spring in Latium and of a pool in the Forum. Vergil makes her the sister of Turnus, Aeneas' antagonist. She was worshipped by all who used water in their work. The Dioscuri (cf. January 27) are said to have appeared at her pool after the battle of Lake Regillus in 497 B.C.; and their association with Juturna appears to be of ancient date: it is possible that both were introduced from Ardea which was under strong Etruscan, and hence Greek, influence. The name Juturna or Diuturna has been derived from the Latin root found in Jupiter and from an Etruscan suffix denoting filiation: this would be a rare instance of such relationships in native Roman mythology.

Temple of Janus: in 29 B.C. Octavian closed the gates of the temple of Janus, indicating the restoration of peace after the battle of Actium. The temple had been closed only twice before, once by Numa and once after the first Punic war in 235 B.C. The shrine of Janus was a free-standing double arch or *ianus geminus* in the Forum, through which perhaps Roman armies left for battle. The god may have been conceived as two-faced since the arch looked both ways. Janus has been explained as Dianus, a male counterpart of Diana, and thus a sky-god like Jupiter; or as the *numen* of doors, *ianuae*, gates, or bridges and thus as a god of entrances and beginnings: hence the name January for the first month of the Roman year. Since his priest was the *rex sacrorum*, Janus may have been the particular *numen* of the king's door in the *regia*, and consequently elevated to great importance: he was addressed before Jupiter in prayers.

13. *Augustus* represented his new Principate as a restoration of the Roman Republic, and for this restoration he was awarded an oak crown by the Senate.

    Iovi Statori circenses: circus games for Jupiter Stator (cf. June 27; September 5 and 23). They were begun when a consul dropped a *mappa* or napkin (cf. January 7 and April 19).

14. *Mark Antony* born, 82 B.C. From the time of Augustus to the reign of Claudius, Antony's birthday was held *vitiosus*, "full of corruption."

    Nero Claudius Drusus was born to Livia about the time

of her marriage to Augustus in 38 B.C. He and his descendants profited from the suspicion that he might be a son of Augustus and not of Tiberius Claudius Nero, Livia's former husband.

15. *CARMENTALIA:* cf. January 11. According to Ovid, this second festival of Carmenta was added after a Roman army left by Carmenta's Gate to attack Fidenae and returned victorious, or established by Roman matrons angered at the rescinding of their privilege to ride in carts called *carpenta*. It is probably a continuation or repetition of the rites of January 11.

16. *Augustus:* this title was conferred on Octavian by the Senate in 27 B.C. The word is connected with *augere*, "to increase," and suggests the enhancing of dignity and power.

Concordiae Augustae: a temple to Augustan Harmony was dedicated by Tiberius in 10 A.D. Cf. July 22.

24. *Sementivae:* rites of sowing, perhaps identical with the Paganalia or Country-Rites. Ceres and Tellus were invoked to protect the seed, the plough-oxen were garlanded, and there were offerings of cakes and the sacrifice of a pregnant sow.

Hadrian, born on this day in 76 A.D., succeeded Trajan on the throne in 117 A.D. and ruled until his own death in 138 A.D. He was a great traveller and a lover of Athens which he embellished with various monuments, including a great library. His large villa at Tibur and his mausoleum in Rome are also conspicuous monuments and worthy memorials of an excellent reign.

27. *Castori et Polluci:* the cult of Castor and Pollux, brothers of Helen, was introduced at Rome after the Dioscuri were thought to have aided the Romans against the Latins in the battle of Lake Regillus. This Greek cult may have been imported from Ardea or Tusculum (cf. January 11). A temple to the Dioscuri was dedicated in the Forum in 484 B.C. and restored by Tiberius in 6 A.D. January 27 is the date of dedication or of re-dedication; this is uncertain. Remains of this temple are a conspicuous monument in the Forum today. The *Castores*, as the Romans usually called them, were special patrons of the knights or *equites*.

30. *Ara Pacis:* this great Altar of Peace was dedicated by Augustus in 1 B.C. Restored at the present day, it is properly regarded as one of the masterpieces of Augustan art.

Livia, wife of Augustus, was instrumental in securing the succession for her son, Tiberius. Throughout a long life, she embarrassed him with a sense of this obligation; and when she died in 29 A.D. at the age of eighty-plus years, Tiberius paid her no homage.

31. *Antonia Minor,* the youngest daughter of Antony and Octavia, married Drusus, Tiberius' popular brother, and became by him the mother of Germanicus and of the future emperor, Claudius. She was famous for her beauty and integrity. She died in 38 A.D. shortly after the accession of her monstrous grandson, Caligula.

FEBRUARY

1. *Iunoni Sospitae:* a temple to Juno the Savior—Sospita, Sispes, or Seispes—was dedicated in the vegetable market, the *forum holitorium,* in 197 B.C. This cult may have been introduced from Lanuvium in Latium where Juno was also worshipped as Sospita. Her image there was clad in a goat-skin coat. Juno was worshipped in Rome as early as 509 B.C. when the great Capitoline temple was dedicated to Jupiter, Juno, and Minerva; and the cult of this triad was probably formalized under Etruscan influence long before that. Juno's name has been derived from the *iuno* which every woman possessed, as each man possessed his *genius,* or from *Diuno* bearing the same root which appears in the names of Jupiter, Diana, and perhaps Janus. The Kalends of every month was sacred to Juno; and in the announcement of the Nones, which gave the Kalends its name, she was addressed five or seven times as Juno Covella to indicate whether the Nones would fall on the fifth or seventh day of the month. This ceremony certainly connects Juno with the moon either as a goddess of light or as a goddess of women to whom the lunation is important because of the menstrual cycle. In any case, she is an old Italian goddess later identified with the Greek Hera, though her nearest linguistic counterpart in Greece is probably Dione. Juno probably gave her name to the month of June in an Etruscanized form, i.e. Junius, as if from Uni, instead of Junonius. Ovid, however, gives some credence to a derivation of June from *iuniores,* "the younger."

5. *Concordiae in arce:* a temple to Harmony on the citadel dates back to 216 B.C.

Pater patriae: Augustus was named father of his country in 2 B.C.

13. *Parentalia:* rites for the dead, especially deceased parents and other ancestors, begun on this day with a sacrifice at the tomb of Tarpeia by one of the Vestal virgins. During this period of nine days temples were closed, marriages were forbidden, and the magistrates appeared without insignia of office.

Fauno in insula: rites celebrating the founding of Faunus' temple on the Tiber island in 196 B.C. It was financed by fines exacted from holders of the *ager publicus* who had not paid their land-rents for using the public domain. Faunus has been explained as a wolf-god connected with the Lupercalia of February 15, or as a propitious deity whose name is derived from *favere.* He bore the titles Fatuus and Fatuclus meaning "speaker" and implying oracular powers. In antiquity he was identified with Inuus, a fertility god, with Incubo, a *numen* of nightmares, and with the Greek god Pan. He had a female counterpart, Fauna; and both masculine and feminine forms appear in the plural as a rough equivalent for satyrs and nymphs of Greek mythology.

15. *LUPERCALIA:* fertility rites begun at the Lupercal where Romulus and Remus were said to have been suckled by the *lupa* or she-wolf. Goats and dogs were sacrificed to Faunus or Inuus; and Luperci or Creppi, young men dressed in goat-skins, ran around the base of the Palatine hill: with goat-hide thongs, called *amicula Iunonis* or "mantles of Juno," they lashed out at women whom they met; and the touch of the lash was thought to impart fertility. The word Luperci has been interpreted as "wolf-people"; "averters of wolves" (from *lupus* and *arcere*); or as "wolf-goats" (from *lupus* and *hircus*). The word Creppi may come from *caper,* another term for male goat. These wolf-priests or goat-priests, as the case may be, have been represented as drawing a magic circle around the early Palatine city (cf. the Ambarvalia of May 29). The Lupercalia may have been, in primitive times, the beginning of a breeding cycle for sheep and goats (cf. April 21 and July 7), which, on a 4-month calendar, terminated with the Liberalia of March (cf. March 17). There is some evidence that Liber, the god of the Liberalia, was also associated with the Lupercalia. In a 4-month year, the Lupercalia would have fallen in June. It was at the Lupercalia in 44 B.C. that Antony, himself a Lupercus, offered Caesar a crown.

17. *QUIRINALIA:* the rites of Quirinus, in legend the Sabine equivalent of Romulus. These ceremonies took place in

Quirinus' temple on the Quirinal hill, and the *flamen Quirinalis* presided. He was a major *flamen*, like those for Jupiter and Mars, and unlike most *flamines*, he had religious duties concerned with other gods than his own: he sacrificed to Robigus on April 25, to Consus on August 23, and to Acca Larentia on December 23. Quirinus was mentioned in prayers after Jupiter and Mars, and like them he could receive *spolia opima*, the rich spoils of an enemy dedicated by a Roman conqueror. Quirinus may have been a war-god of the Quirinal hill, as Mars was the war-god of the Palatine hill: the sacred arms of Quirinus were anointed by the *flamen Portunalis*. The name Quirinus has been explained as Sabine in origin, either from *co-viri-um* meaning "an assembly of men"; from *curis*, a spear; from the name Cures, a Sabine town; or from Quirium, presumably a settlement on the Quirinal hill.

Fornacalia: the feast of ovens which ended on this day with a public banquet honoring the goddess Fornax or Oven. It began several days earlier with offerings of meal-cakes in each *curia* or ward of the city. There were thirty *curiae* in all.

Stultorum feriae: the feast of fools, so called because they had neglected to perform the rites of Fornax on the day assigned for such ceremonies in their own *curia*. They might compensate for their "folly" by performing the city-wide rites on this day.

21. *FERALIA:* rites to appease the Manes (the "good ones"?) or souls of the dead. *Epulae* or dinners were carried to the tombs of the dead; hence the word Feralia is commonly derived from *inferi*, "the dead" and/or *ferre*, "to carry."

22. *Caristia:* a family festival for the "dear ones," a sort of reunion of the dead and the living before the Lares of the household.

23. *TERMINALIA:* sacrifices to Terminus, the god of boundaries, on the last day of the old Roman year; thus he may be regarded as a god of terminations, both spatial and temporal. Terminus was represented by a stone in the great Capitoline temple from which, it was said, he would not move when Jupiter was introduced. There was a hole in the roof over this stone because Terminus had to be in the open air. It is possible that this stone was a boundary stone of the early city. In the country neighbors met at their dividing boundary lines on this day to sacrifice a lamb and a suckling pig to Terminus.

24. *REGIFUGIUM:* this is one of only two early Roman festivals occurring on even-numbered days (the other being the Equirria of March 14)—attesting the strength of a superstition against even numbers. The Regifugium or Flight of the King was thought in antiquity to celebrate the expulsion of Tarquinius Superbus from Rome, but modern scholars are prone to regard it as a vestige of the expulsion of a mock king who may have "ruled" the intercalary period which was commonly inserted between the Terminalia and the Regifugium. The *rex sacrorum* may have represented this mock king in Regifugia of the historical period. There may be a parallel to the sequence Terminalia-Regifugium in the sequence Tubilustria-Q.R. C.F.-days (cf. March 23 and 24 and May 23 and 24) : this would imply that the king entered the *comitium* on the first day in each sequence and left it on the last. The following analogies involve a violent exit for the king at the end of similar periods: at Diana's shrine near Aricia a fugitive slave made himself *rex nemorensis* or "king of the grove" by slaying his predecessor in this office; there is a story that Romulus was slain by angry senators at the original Poplifugia (cf. July 5) ; and the prince of the Saturnalia may have been slain or violently expelled in early times.

27. *EQUIRRIA:* horse races on the Campus Martius in honor of Mars. Cf. March 14.

Constantine the Great laid claim to the throne at the death of his father, Constantius, in 306 A.D. He defeated Maxentius, his chief rival, at the battle of the Milvian bridge in 312 A.D. It was in this battle that he embraced Christianity. Constantine moved the capital of the Roman Empire from Rome to Byzantium which he rebuilt and renamed Constantinople. He died in 337 A.D. shortly after his baptism as a Christian. A relic of his reign is the Arch of Constantine in Rome.

MARCH

1. *Feriae Marti:* New Year's Day in the early calendars, sacred to Mars with whose month the year originally began. The sacred fire of Vesta was rebuilt, and fresh laurel boughs were hung on the *regia*. On this day the original *ancile* or sacred shield of Mars was said to have fallen from heaven in the time of Numa. Eleven copies of this shield were made by a smith whose name, according to the story, was Mamurius Veturius. The *ancilia* were

shaped like a figure 8, and they may have been regarded as
actual embodiments of Mars. On March 1, 9, and 23 these
shields were moved by the Salii or Leaping Priests, perhaps
from Mars' sanctuary in the *regia* to various points in the
city. Clothed in *trabeae* or purple-striped robes and *tunicae
pictae* or embroidered tunics, the Salii made procession
through the streets—leaping, dancing, and beating on the
ancilia with staffs. They were twenty-four in number,
twelve Palatini and twelve Collini, Agonenses, or Agonales,
worshipping Mars Gradivus, the Marcher, and Quirinus
respectively. They had to be of patrician birth and have
both parents living. Their dance has been explained as a
war dance or as an exorcism of evil spirits at the beginning
of the new year.

Mars was worshipped in other parts of Italy, but his cult
may have been introduced from Rome: he is usually regarded as a war-god of the Palatine hill, as Quirinus was
the war-god of the Quirinal hill. But in the Hymn of the
Arval Brothers he is addressed as an agricultural deity:
this may point to a chthonic origin which associated him
with death and consequently war. In Rome variants of his
name were Mavors, Marmar, Marmars, and Mamers; in
Etruria he was known as Maris. The name, in any of its
formations, has been derived from *mar*, "to shine," from
mas, "male," or from *mors*, "death."

Mars was paired or otherwise associated with the female
deities Bellona and Nerio or Nerine; but through an identification with the Greek god, Ares, he is often represented
as the lover of Venus. Ovid describes his marriage to Anna
Perenna (cf. March 15), an old woman who substituted
for Minerva in this ceremony. In Roman legend Mars was
the father of Romulus and Remus. The wolf and the wood-
pecker were sacred to him.

Iunoni Lucinae: rites celebrating the anniversary of Juno
Lucina's temple on the Esquiline hill. This was founded in
the fourth century B.C. As Lucina, Juno was a goddess of
child-birth, worshipped especially by matrons: they received gifts from their husbands on this day, and they
entertained their husbands' slaves at dinner. Diana also
bore the name Lucina in certain rites.

Matronalia: another name for the festivities honoring Juno
Lucina. In these rites matrons took a prominent part.

6. *Augustus* was made *pontifex maximus* on the death of
Lepidus the Triumvir in 12 B.C.

7. *Vediovi:* rites to Vediovis *inter duos lucos*, "between two

groves," on the Capitoline hill. This was the site of an ancient asylum, but its connection with Vediovis is uncertain. These rites for Vediovis may be the original ones, constituting the very first festival of the 4-month year, and combining with the rites of the Caprotine Nones which later got separated and fell in July (cf. July 7). Cf. January 1 and May 21.

9. *Arma ancilia movent:* for the second time in this month (cf. March 1), the Salii moved the sacred shields of Mars, here called "arms and shields" or "shield-arms." They were moved again on March 23.

10. *Tiberius* was made *pontifex maximus* after his accession to the throne upon the death of Augustus in 14 A.D.

14. *EQUIRRIA:* horse races in honor of Mars. Cf. February 27. Contrary to the Roman superstition, this event occurs on an even-numbered day. See note on February 24.

Feriae Marti: rites probably identical with those of the Mamuralia described below.

Mamuralia: the festival of Mamurius Veturius, the fabled smith of the *ancilia* (cf. March 1). A man clothed in skins and reviled as Mamurius was driven out of the city because the shields of Mamurius, so it was said, brought ill fortune to the Romans. The Salii may have had something to do with this ceremony. The name Mamurius Veturius is now thought to be a variant of Mars, meaning "Old Mars," i.e. "Mars of the Old Year." The expulsion of Mamurius may be regarded as the expulsion of the old year or as the expulsion of a *pharmakos* or scape-goat on which old afflictions are loaded in order to be rid of them.

15. *Annae Perennae:* festival of the "unending moon-cycle," if *anna* is a feminine form of *annus*. The name has also been explained as an abstraction of the verb *annare*, "to pass through a year"; as a form of *anus*, "old woman"; or as a corruption of Perna, the name of a goddess with Etruscan connections and perhaps an equivalent of Ceres. Ovid identifies Anna Perenna with Dido's sister who is said to have wandered to Italy and, after visiting Aeneas and Lavinia, ended her life in the river Numicius. Anna is also represented as an old woman of Bovillae who brought *liba* or cakes to the famished plebs when they seceded to the Sacred Mount; or again as an old hag who substituted for Minerva in marriage-rites with Mars. The festival of Anna Perenna was held at the first mile-stone on the Via Flaminia: tents were pitched near the Tiber,

and celebrants spent the day in revelry and drinking. It is possible that these rites represent an original New Year's celebration at the appearance of the first full moon of the new year.

Caesar was assassinated by the Liberators on this day in 44 B.C. The festival of Anna may have created a useful distraction.

16. *Sacra Argeorum:* processions to the Argei, 24 or 27 places where shrines of some sort existed. They may be connected with the straw figures, also called Argei, which were thrown into the Tiber by the Vestal Virgins on May 15 (cf. May 15).

17. *LIBERALIA:* rites for Liber, an old Italian fertility god later identified with Dionysus, the Greek god of the vine. Old women sold *liba* or cakes to the celebrants; and there may be some connection between this word and the god's name. Liber is sometimes identified with Jupiter (cf. September 1) who also had functions related to viticulture (cf. April 23). Perhaps through a verbal misconception, this was also the day on which boys, in early times, assumed the *toga libera* or *virilis*, the cloak of freedom or manhood. Originally, the Liberalia may have terminated a breeding cycle for goats and sheep, which began at the Lupercalia (cf. February 15). In that case, Liber is probably the *Indiges* (cf. January 9) of the *agonalia* mentioned below, and a god of copulation not unlike Inuus (cf. January 9) and perhaps Consus (cf. August 21). Cf. April 19.

AGONALIA: here perhaps a priestly term for the festivities of the Liberalia, called Agonalia by the Salii Agonenses (cf. January 9, May 21, and December 11).

19. *QUINQUATRUS:* so named because this was the fifth day after the Ides. This day was regarded as the birthday of Minerva who is thought by some to be the equivalent of Nerio or Nerine, the Sabine goddess, or of Bellona, both of whom were associated with Mars, perhaps as his wife. However, Minerva's connection with the Quinquatrus may be accidental, resulting from the coincidence of a temple-dedication to her on this day: she had one shrine on the Caelian hill and another on the Aventine. It appears that she was first introduced to Rome as a member of the Capitoline triad (an Etruscan arrangement), and she has been explained simply as the Greek Athena as known in Etruria. Others regard her as an Italian goddess of handicrafts whose name is derived from the root found in

meminisse, "to remember," and in *mens,* "mind." Through confusion over the meaning of Quinquatrus, the term was interpreted by the Romans to authorize a five-day festival for Minerva and a five-day vacation for school-boys. Cf. June 13.

Feriae Marti: more rites for Mars. The Salii danced in the *comitium* or assembly-place and purified the *ancilia* or sacred shields (cf. March 1). The *tribuni* of the three ancient tribes also participated in these rites which some regard as the original Quinquatrus.

23. *TUBILUSTRIUM:* the purification of the *tubi* or *tubae,* perhaps the trumpets used to summon the assembly on the following day, or to mark the beginning of the campaigning season. For some reason these rites took place *in atrio Sutorio,* in the Shoemaker's Hall. Cf. May 23 and October 19. On this day the *ancilia* were moved again. Cf. March 1 and 9.

24. *Q.R.C.F.:* "Quando rex comitiavit, fas"—"When the king has summoned the assembly, it is right" to perform legal business. It is thought that the *comitia* sanctioned wills on days marked thus, i.e. March 24 and May 24, and that the *rex sacrorum* was somehow involved in this process. However, there was an ancient belief that the letters stood for "Quod rex comitio fugerat"—"Because the king had fled the assembly," referring to the expulsion of Tarquinius Superbus, the last king of Rome. This phrase, if authentic, may have referred to the expulsion of a mock king (cf. February 24). Cf. May 24.

25. *Hilaria:* "joyful rites" of Cybele, celebrated on the first day after the vernal equinox with a procession and masquerades. Actually this was the continuation of rites begun about March 22 by the Galli, priests of Cybele. Cybele was the Anatolian mother-goddess whose cult was introduced at Rome about 200 B.C. Her temple was on the Palatine hill, and her priests were Oriental until the time of Claudius in the first century A.D. The temple rites which preceded this public festival of the Hilaria commemorated the castration and death of Cybele's youthful lover, the vegetation god Attis. Cf. March 27 and April 4.

27. *Lavatio:* more rites for Cybele, which involved bathing the goddess' image in the river Almo.

Caesar recovered Alexandria from Cleopatra's enemies in 47 B.C.

30. *Iano, Concordiae, Saluti, et Paci:* Ovid mentions only the need of worshipping Janus, Harmony, Safety, and Peace on this day.

31. *Lunae in Aventino:* rites to the Moon, i.e. Diana, on the Aventine hill.

APRIL

1. *Veneralia:* rites for Venus Verticordia, "turner of hearts," worshipped on this day by women of rank. Her temple was founded in 114 B.C. Venus was an old Italian goddess, perhaps, as partner of Mars, equivalent to the Sabine deity Nerio or Nerine. Venus was later identified with the Greek Aphrodite; and as Aphrodite, she was thought by some to have given her name to the month of April. Ovid gives another derivation of April from *aperire,* "to open," implying that this is the month when buds open and nature releases her bounty. A recent view makes April the month of the "boar," *aper.* It has been pointed out that Venus' name ought to be neuter, referring perhaps to some charming thing, e.g. a vegetation plot. Lucretius seems to retain some reminiscence of Venus' powers as an earth goddess. The verb *venerari,* "to venerate," may be formed on her name, attesting the great extension of religious feeling inspired by her rites. As the traditional mother of Aeneas' race and of the Julian clan, she received much attention from the early Caesars. As a goddess of love, she had great popularity as late as the fourth century A.D. Cf. April 23, August 19. and September 26.

   Fortunae virili: rites for Manly Fortune were conducted by women of the lower classes in the men's baths. Fortuna was asked to bless them in their relations with men. Fortuna appears to be an agricultural deity introduced from Antium or Praeneste. Her name, derived from *ferre,* implies that she was a "bringer" of luck or, even before that, of fertility. As *primigenia* she was said to be the "first-born" daughter of Jupiter, or indeed the primal mother of Jupiter and Juno. Fortuna was sometimes identified with the Etruscan goddess, Nortia. Cf. November 13.

4. *Magnae Matri:* rites for Cybele, the Great Mother of Mt. Ida. In 204 B.C. the Romans, in accord with the Sibylline Books, imported from Phrygia a stone representing this goddess. On this day, April 4, the stone was deposited in the temple of Victory on the Palatine hill. The Great

Mother was thought to have brought the Romans success in concluding the war against Hannibal. Cf. March 25 and 27, and April 10.

Ludi Megalesiaci: the Megalesian Games in honor of the Great Mother. Romans were forbidden to serve the goddess, but they celebrated with dinner-parties.

5. *Fortunae publicae:* rites celebrating the founding of the temple to Public Fortune on the Quirinal hill. Cf. May 25.

6. *Thapsus:* Julius Caesar defeated the Republican forces and their ally, King Juba, on this day in 46 B.C. It was shortly after this battle that Cato the Younger committed suicide at Utica, rather than surrender to Caesar.

10. *Magnae Matri:* a temple to the Great Mother on the Palatine hill was dedicated on this day in 191 B.C., and its anniversary marked the last day of the Megalesia. Cf. March 25 and 27, and April 4.

11. *Septimius Severus* was born in Africa of Syrian stock in 146 A.D. He was proclaimed emperor in 193 A.D. and ruled until his death at York in England in 211 A.D. Most of his reign was consumed in wars against his enemies, Roman and barbaric. He reorganized the Roman army on a democratic basis, making it the source of his power and the chief object of his attention. His Arch is still a conspicuous monument in the Forum.

12. *Ludi Cereales:* the Games in honor of Ceres began on this day and terminated with the Cerealia on April 19. Cf. April 19.

13. *Iovi Victori:* a temple to Victorious Jupiter was dedicated by Quintus Fabius Maximus in 295 B.C.

15. *FORDICIDIA:* the slaughter of *fordae* or *hordae,* pregnant cows, one on the Capitoline hill, and others in each of the thirty *curiae* or wards of the city. The cows were sacrificed to Tellus, the Earth; the unborn calves were removed by the chief Vestal Virgin and burned: the ashes were kept by the Vestals for use at the Parilia (cf. April 21). The rites of the Fordicidia were clearly fertility rites of some kind.

16. *Dies imperii:* on this day in 29 B.C. Octavian took the title *imperator,* "general" and later "emperor," so it was henceforth regarded as the birthday of the Roman Empire.

19. *CEREALIA:* the rites of Ceres, an ancient Italian deity later identified with Tellus and with the Greek Demeter.

Her name is connected with the verb *creare*, "to create," and with an obsolete word *cerus*, equivalent to classical Latin *genius*, the term for something which every man possessed. It has therefore been suggested that Ceres was a feminine form of *genius*, perhaps another name for the *iuno* possessed by every woman. Under Greek influence Ceres had a formidable aspect: a mad person was called *larvatus*, "possessed by a ghost," or *cerritus*, "possessed by Ceres." For some reason it was the practice at the Cerealia to fasten burning brands to the tails of foxes, and then to release the animals in the Circus Maximus. This may have had something to do with the purification of crops, since Ceres was primarily the goddess of grain. There was a *flamen Cerealis*. Cf. April 12.

Ludi circenses: more games proclaimed by a consul who dropped a *mappa* or napkin to signal the start (cf. January 7 and 13) of the races.

Cereri, Libero, et Liberae: a temple to Ceres, Liber, and Libera was founded in 496 B.C. to dispel a famine; it was dedicated on April 19, 493 B.C. It was built at the foot of the Aventine hill and financed by fines imposed by the tribunes of the people. Once built, the temple was supervised by the plebeian aediles. For Liber, see the note on March 17. Libera is his female counterpart, later identified with the Greek Persephone, and thus regarded as Ceres' daughter.

21. *PARILIA:* or Palilia, the rites of Pales, a deity whose name is either masculine or feminine in form and either singular or dual in number (cf. July 7). In the country sheep-folds were decked with boughs: at dawn the shepherd purified his sheep and the sheep-fold, then made a fire through which he leaped. In Rome a mixture of blood from the October Horse (cf. October 15) and ashes from the Fordicidia (cf. April 15) was thrown upon a fire of bean-straw on the Palatine hill, and celebrants leaped through the flames. These rites of purification were regarded as somehow connected with the founding of Rome, and this day was celebrated as the birthday of the city and, in some sense, as the beginning of a new year. The Parilia may be a copy of the rites of the Caprotine Nones (cf. July 7), and originally the beginning of a breeding cycle for goats and sheep in a 4-month calendar. This cycle would have terminated in the *agonalia* of May (cf. May 21), and the Caprotine Nones would have been just halfway through this cycle. This would account for the apparently odd date of the Parilia and for its celebration as the beginning of

a year. In a 4-month year, the Parilia would have been
concurrent with the Consualia of August 21 and with the
Divalia of December 21, so those festivals may be dislodged
portions of the Parilia. Pales has been explained as a god
of the Palatine hill, as an Etruscan importation, or as
a god of stock-breeding introduced by northern invaders.
For the two Pales as twin kids, cf. July 7. There was a
flamen Palatualis and, on December 11, an offering called
the Palatuar.

23. *VINALIA:* a wine festival sacred to Venus, Jupiter, or
both at different periods in history. According to Ovid, the
temple of Venus Erycina, i.e. of Mount Eryx in Sicily,
was established in Rome on this day; and Livy gives the
date as 181 B.C. Libations of new wine were poured to
Jupiter at the temple of Venus. Cf. August 19.

24. *Feriae Latinae:* the Latin rites for Jupiter Latiaris began
about this time on the Alban Mount; the exact date was
announced by the consuls when they took office. The
consuls attended these rites in person, and they were not
permitted to leave for their provinces until these rites were
completed. This *Latiar,* as it was called, was an ancient
festival of the Latin towns, during which peace prevailed
among warring cities and Rome renewed her alliance with
Latin neighbors. The Roman consuls slaughtered a white
heifer, and the meat was divided among representatives
from all the towns.

Jupiter is a Latin version of the Indo-European sky-god:
the root of his name appears in Sanskrit Dyaus, in Greek
Zeus, and possibly in Janus. It also appears in the names
of the goddesses Dione, Diana, and perhaps Juno. In Rome
the Ides of each month was sacred to Jupiter; and the *flaminica Dialis* sacrificed to him on all *nundinae* or market-
days. Along with Mars and Quirinus, Jupiter received
spolia opima, "rich spoils," and possessed a major *flamen.*
He was the central figure in the Capitoline triad, formalized
under Etruscan influence in the sixth century B.C. Roman
triumphal processions always terminated at his temple on
the Capitoline hill. Banquets or *epula* were served to
Jupiter on September 13 and November 13.

25. *ROBIGALIA:* rites for Robigus or Robigo, Rust or Mildew,
performed by the *flamen Quirinalis* at the fifth mile-stone
on the Via Claudia. A red dog and a sheep were sacrificed
to Robigus to avert blight from the grain. Robigus had
some connection with Mars.

26. *Marcus Aurelius,* the philosopher prince, was born on this

day in 121 A.D. He succeeded his adoptive father, Antoninus Pius, on the throne in 161 A.D. and ruled until his death in Pannonia in 180 A.D. A column, still extant in Rome, celebrates his victories over the Germans. He was a follower of the Stoic philosophy, but his creed did not prevent persecution of the Christians. His *Meditations* in Greek record his casual reflections on life.

28. *Ludi Florae:* games for Flora, an Italian vegetation deity. They were established to end a dearth in 238 B.C. They were supervised by the plebeian aediles. A temple to Flora in or near the Circus Maximus was also dedicated on this day in the same year. Like the temple of Faunus on the Tiber Island (cf. February 13), this temple of Flora was probably financed with fines exacted from delinquent renters of the *ager publicus.* Flora had another temple in the Sabine quarter of Rome, and there was a *flamen Floralis.* Flora is one of the deities mentioned in the Acts of the Arval Brothers. The rites of Flora, celebrated with some indecency, were observed especially by prostitutes. Hares and goats were released in the Circus Maximus; and vetches, beans, and lupines were scattered among the people in the Circus. These festivities continued through May 3.

MAY

1. *Ludi Florae:* see note on April 28, when these Games began.

Bonae Deae: rites to the Good Goddess, celebrating the anniversary of the dedication of her temple on the Aventine hill. This temple was later restored by Livia, the wife of Augustus. The rites of May 1, unlike those of December 3 (cf. December 3), were held in this temple and not in the house of a priest or a magistrate. Men were excluded from both rites. Wine could be brought into the temple only under the name of milk, and it had to be carried in a honey-jar or *mellarium.* Herbs were kept in the temple, but myrtle was specifically excluded. The sacrificial victim was a pig, and it was called Damium. The goddess herself was called Damia and her priest a Damiatrix. The *flamen Volcanalis* was somehow involved in these ceremonies. Bona Dea was sometimes identified with Maia, Fauna and Ops.

Laribus praestitibus: this day was sacred to the *Lares praestites,* the protecting Lares, to whom an altar and statuettes were erected. The statuettes were dressed in dog skins; the figure of a dog was put at their feet; and probably the victim of sacrifice was a dog. (Cf. April 25.)

7. *Mola salsa:* from May 7 to 14 the Vestal virgins made the salted meal for use at the Vestalia on June 9, at Jupiter's Banquet on September 13, and at the Lupercalia on February 15. It was made from the first ears of new grain by the three senior Vestals. An iron saw was used in preparation of the salt: this may indicate an Iron Age origin for the cult of Vesta, goddess of the hearth-fire. The Penates should be associated with her as gods of the *penus* or store-room. It has been suggested that Vesta and the fire-god Vulcan were preceded by Cacus and Caca, two fire-gods of the Palatine hill.

9. *LEMURIA:* rites of the dead occurring on three alternate days, May 9, 11 and 13. (Cf. February 13, 21 and 22.) The Lemuria were said to have been established to expiate the death of Remus. The *lemures* appear to have been unfriendly ghosts or *larvae* while the *manes* were friendly spirits. In historical times the Lemuria were domestic rites performed by the *pater familias,* the head of the family. He walked through his house at midnight, spitting out black beans which the *lemures* of his ancestors were supposed to gather up.

11. *LEMURIA:* see note on May 9.

12. *Ludi Marti in circo:* games to Mars in the Circus.

Marti Ultori: a temple to Mars Ultor, the Avenger, was vowed by Octavian after the battle of Philippi in 42 B.C., but it was not dedicated until May 12 or August 1, 2 B.C.

13. *LEMURIA:* see note on May 9.

14. *Marti invicto:* nothing is known about these rites for the unconquered Mars.

15. *Sacra Argeorum:* rites of the *Argei,* straw puppets cast into the Tiber from the old wooden bridge, the *pons sublicius.* They were brought to the bridge by the *pontifices* ("bridge-builders"?) and thrown into the river by the Vestal virgins. The *flaminica Dialis,* the priestess of Jupiter, wore mourning garb at this ceremony. The puppets were thought to represent Greeks, i.e. Argives, who had followed Hercules to Italy, or old men who had once been sacrificed in the Tiber. The rites were performed in honor of Saturn or Dis Pater, i.e. Pluto, according to ancient authorities. They are explained today as vegetation rites, as an expulsion of demons, or as expiations of the river-

god, Tiber. Ovid, perhaps rightly, assigns these rites to May 14. Cf. March 16.

Feriae Iovi, Mercurio, Maiae: rites to Jupiter, Mercury, and Maia but largely devoted to the worship of Mercury, the god of trade. Jupiter and Maia were included because they were regarded as his parents through identification with Greek counterparts. Moreover, every Ides was sacred to Jupiter. Mercury has been explained simply as the Greek Hermes with a Latin name, or as a Greco-Etruscan deity with functions of the Latin *genius* which every male possessed. Maia was identified in antiquity with Tellus and Bona Dea, and confused with the Greek Maia, daughter of Atlas and mother of Hermes. Her name has been derived from the root *mag-* found in *magna*, "big," suggesting that she was a goddess of growth or increase; from the root *ag-* which appears in *aio*, "I speak," and in *magia*, "speaker," hence "wise-woman," and from **maia* meaning "sow"—a root found in *maialis*, "gelding boar." At Tusculum Jupiter bore a masculine form of Maia's name, i.e. Maius. Maia was also known as Maiesta, and she was sometimes regarded as a cult partner of Vulcan. She may have given her name to the month of May, though Ovid gives some credence to a derivation from *maiores*, "the elders."
The first temple of Mercury was dedicated in the Circus Maximus on this day in 495 B.C.

21. *AGONALIA:* see notes on January 9, March 17, and December 11. The *agonalia* of May 21 may have terminated a breeding cycle for goats and sheep which began at the Parilia (cf. April 21) in a primitive 4-month year.

Vediovi: see notes on January 1, March 7, and July 7.

23. *TUBILUSTRIUM:* see note on March 23.

Volcano: rites to Vulcan, the fire-god. He was supposed to have made the *tubae* which were purified on this day (cf. March 23). The *flamen Volcanalis* sacrificed to Bona Dea on May 1, and Maia was sometimes addressed as Vulcan's Maia. Vulcan is explained as an Italian volcanic fire-god, in Rome a successor to Cacus, or as the Greek Hephaestus introduced from Etruria. Vulcan was also called Mulciber, the god who "soothes," i.e. quenches fire; and this may account for his association with water: fish were sacrificed to him on August 23, a day on which Juturna and other water nymphs received attention; and Vulcan was an important deity at Ostia, Rome's port. Cf. August 23.

24. *Q.R.C.F.:* see note on March 24.

Germanicus, nephew of Tiberius and husband of Augustus' grand-daughter, the Elder Agrippina, was born on this day in 15 B.C. He was adopted by his uncle on the latter's accession to the throne in 14 A.D., and served him as supreme commander in Germany and then in the East. He quelled a mutiny in the Rhine armies and undertook expeditions into Germany, but Tiberius suspected his loyalty and policy. He died at Antioch in 19 A.D. and was thought to have been poisoned by Gnaeus Piso, an agent of Tiberius. Among his children were the future emperor, Caligula, and Agrippina the Younger, mother of Nero.

25. *Fortunae publicae populi Romani:* rites to the Public Fortune of the Roman people. One of three temples to Public Fortune on the Quirinal hill was vowed in 204 B.C. and dedicated ten years later: this day may be an anniversary of the dedication. Fortuna may be identical with Fortuna Primigenia, the First-born, a goddess introduced from Praeneste. Cf. April 5.

29. *Ambarvalia:* a religious procession which moved three times around the *arva* or ploughed fields; it may have been led by the Arval Brothers, the priests of the ploughed fields. It probably consisted of *suovetaurilia,* i.e. the driving around and sacrifice of a pig, a sheep, and a bull. The rites were performed in honor of Mars or Ceres, and seem to represent the drawing of a magic circle around the fields. Cf. the Lupercalia of February 15.

JUNE

1. *Iunoni monetae:* the temple of Juno the Warner (*moneta* from *monere?*) was dedicated in 344 B.C. on the *arx* or citadel of the Capitoline hill; but her precinct there was probably older: her sacred geese were said to have wakened the Roman sentries and saved the Capitol from the Gauls in 390 B.C. Part of the temple of Juno Moneta was used as a mint, hence the developed meaning of *moneta* as mint and the derivation of our word money. Cf. February 1 and October 10.

Carnae: rites to Carna, a deity supposed to ward off *striges* or vampires from infants. Lard and beans were offered to her and eaten by the celebrants to secure a good digestion. Ovid identifies Carna with Cardea, the goddess of hinges or *cardines;* but her name is probably connected with *caro,* "flesh," implying that she was a protectress of the vital organs. She had a shrine on the Velian hill.

Kalendae fabariae: the Kalends of the bean, so named for the beans offered to Carna. This was the season of the bean harvest, but this may be accidental; beans were used in other rites too: cf. the Lemuria of May 9. It has been suggested that June was originally named Fabarius, "month of the bean," and that the Kalends retained this name.

Tempestati: a temple to the Weather was dedicated in 259 B.C.

Marti ad portam Capenam: the temple of Mars at the Porta Capena on the Appian Way was dedicated or re-dedicated on this day.

3. *Bellonae in circo Flaminio:* a temple to Bellona, goddess of battle, was vowed in 296 B.C. and later built in the Circus Flaminius. In front of it was the *columella* or little pillar from which the *fetiales* threw a spear to declare war. This pillar stood in an area purchased, under Roman compulsion, by a captured soldier of King Pyrrhus who had no land in Italy into which a Roman spear could be cast in accordance with the usual practice of the *fetiales.* The temple of Bellona was a customary meeting-place for the senate when it had to sit outside the *pomoerium* or city-limits.

     Bellona or Duellona was an early Roman war-goddess variously identified with the Sabine goddesses Nerio and Vacuna, and later with Ma, the Cappadocian mother-goddess introduced at the time of Sulla's campaigns against Mithridates, 87-83 B.C. The followers of Ma were known as *fanatici,* "people of the *fanum* or shrine."

4. *Herculi magno custodi:* rites to Hercules the Great Guardian held to celebrate the founding of his temple near the Circus Maximus by Sulla in 82 B.C. The cult of Hercules, the Greek Heracles, was introduced to Rome at a very early date, perhaps by traders. Hercules was sometimes identified with the old Italian deity, Semo Sancus Dius Fidius. Cf. June 5 and August 12.

5. *Dio Fidio in colle:* a temple to Dius Fidius was vowed by Tarquin the Proud and later dedicated on the Quirinal hill by Spurius Postumius in 466 B.C. Dius shows the same root as Jupiter, and Fidius occurs both in Umbrian and Oscan dialects, as well as in Latin: there was an Oscan Ides of Fidius, and the word Ides itself was interpreted in Etruria as *Iovis fiducia,* "Jupiter's assurance." As *Iovis*

filius, Dius Fidius was identified with Hercules; and his name was used in an oath, *me dius fidius,* (like that of Hercules, *me hercle*) which had to be taken in the open air.

Dius Fidius was also identified with Semo Sancus or Sangus who had shrines on the Quirinal hill and on the Tiber island. In one of his temples were kept *orbes* or wheels of brass, a statue of Tanaquil, wife of Tarquinius Priscus, and an ancient treaty said to have been struck by Tarquin the Proud with Gabii. There was an augural bird called *sanqualis avis* and a gate on the Quirinal hill called the *porta sanqualis.* The name *Sancus,* often misconstrued as *sanctus,* "holy," has been derived from *sancire* in its basic sense of "strike," as in the case both of lightning and a treaty. Sancus was served by Bidental priests who tended places struck by lightning. *Semones* are addressed in the Hymn of the Arval Brothers, apparently as gods of sowing.

7. *Vesta aperitur:* "Vesta is open." On this day the *penus Vestae* or storehouse of Vesta was opened to Roman matrons who entered it barefoot to begin the rites of the Vestalia (cf. June 9).

Ludi piscatorii: the Fishermen's Games.

8. *Menti in Capitolio:* a temple to Mens or Mind on the Capitoline hill was vowed after the battle of Lake Trasimenus in 217 B.C. and dedicated two years later on this day.

9. *VESTALIA:* these rites of Vesta, connected with the first-fruits, began with the opening of Vesta's storehouse on June 7 and terminated on June 15; all days in this interval were both *nefasti* and *religiosi,* i.e. wrong for some legal business and days of ill omen. The *flaminica Dialis,* priestess of Jupiter, might not cut her hair, pare her nails, or have relations with her husband, the *flamen Dialis,* during this period. The Vestals offered cakes of *mola salsa* (cf. May 7); bakers and millers had a holiday, mills were garlanded, and the mill donkeys were garlanded with wreaths and cakes.

Vesta has been explained as a fire-goddess of the king's hearth which was tended by the king's daughters, i.e. the original Vestal virgins. Her cult may have been introduced by the cremating Iron Age race whose burials have been found in the Forum: her round temple reproduces the shape of a burial hut-urn characteristic of Villanovan culture. It has been suggested that Vesta replaced an earlier fire-goddess of the Palatine hill, viz. Caca.

11. *MATRALIA:* rites for Mater Matuta whose temple in the *forum boarium* or cattle market was dedicated in 396 B.C. An earlier structure was attributed to king Servius Tullius. Only wives of a first marriage could deck the goddess' image; and they prayed first for their nieces and nephews, then for their own children. Slaves were not admitted to the temple except for one female who was then driven out during the ceremony. In antiquity Mater Matuta was regarded as a Good Mother, not unlike Bona Dea, or as Mother Morning (as if from *mane*, "morning"). She may have been a primitive goddess of growth whose name Matuta is derived from *matura*, "ripe." She was identified with the Greek Leucothea from whom she acquired a connection with harbors.

Fortunae: a temple to Fortune in the *forum boarium* was said to have been dedicated by Servius Tullius: it contained a draped wooden statue, either of Servius Tullius or of Fortuna herself, which lasted until the time of Sejanus, Tiberius' minister. Sejanus, of Etruscan descent, is said to have owned an ancient statue of Fortuna. Cf. June 24.

Concordiae: a temple to Harmony was dedicated in the Portico of Livia in 7 B.C.

13. *Feriae Iovi:* the usual rites to Jupiter on the Ides of a month; but it appears that a temple to Jupiter Invictus was dedicated on this day in 192 B.C. On this particular day too the *tibicines* or trumpeters feasted in the temple of Jupiter Capitolinus. (See entry below.)

Quinquatrus minusculae: the lesser Quinquatrus which, like the greater Quinquatrus on March 19, involved rites to Minerva, perhaps as patroness of the *tibicines* or trumpeters. (See entry above.)

15. *Q. ST. D. F.:* "Quando stercus delatum, fas"—"When the filth has been carried out, it is right" to do legal business. On this day the rites of the Vestalia were completed by cleaning the temple of Vesta and throwing the refuse into the Tiber or into a receptacle on the Clivus Capitolinus. Cf. June 7 and 9.

18. *Annae sacrum:* a rite for Anna, presumably Anna Perenna (cf. March 15).

19. *Minerva in Aventino:* rites to Minerva, celebrating the anniversary of her temple on the Aventine hill. Cf. March 19.

20. *Summano ad circum maximum:* rites to Summanus in the Circus Maximus. The temple of Summanus was founded in the third century B.C. because lightning had struck a figure of Jupiter on the Capitol. Summanus seems to have been a god of thunderbolts by night, as Jupiter was god of thunderbolts by day. He was thought to be Sabine, but the name may be Latin, i.e. Submanus, "before morning," or simply Summanus, the god of *summa* or high places. Offerings were made in the form of wheel-shaped cakes.

23. *Dies ater:* a "black day" because it was the anniversary of Hannibal's victory over the Romans at Lake Trasimenus in 217 B.C. Hannibal marched through the swamps of the Arno river and surprised the Roman consul, Flaminius, on the banks of Lake Trasimenus. It was on this march that Hannibal lost the sight of one eye. (A different reading of Ovid puts this *dies ater* on June 21.) Cf. July 18 and October 6.

24. *Forti Fortunae trans Tiberim:* rites to Chance Fortune, celebrating the founding of a temple or temples across the Tiber, either by Servius Tullius or by a consul in 293 B.C. These rites were conducted at the first and sixth milestones on the Via Portuensis. Plebeians and slaves were particularly devoted to these rites. Both names, Fors and Fortuna, are derived from *ferre,* "to carry," implying that Fortune was a Bringer of luck. Fortuna Primigenia had an oracle at Praeneste; she was variously depicted with a cornucopia, a ship's rudder, or a wheel. The Etruscan goddess, Nortia, was identified with Fortuna.

27. *Iovi statori:* anniversary of the temple of Jupiter Stator, the Stayer, vowed, so it was said, by Romulus to stay the flight of his men who were retreating before the Sabines. Jupiter Stator may have been simply the Stayer of the calendar, since his rites would have been the last in a 4-month year.

   Laribus publicis in summa sacra via: Augustus built or rebuilt a temple to the public Lares at the top of the Sacred Way. This day was the anniversary of its founding.

29. *Quirino in colle:* anniversary of the temple of Quirinus on the Quirinal hill.

30. *Herculi Musarum:* a temple to Hercules of the Muses was founded in 189 B.C. and restored in the time of Augustus.

JULY

1. *Felicitati in Capitolio:* anniversary of a chapel to Good Fortune on the Capitoline hill.

4. *Ara pacis Augustae constituta:* the great altar of peace was decreed by the senate on July 4, 13 B.C., but not dedicated until January 30, 9 B.C. Cf. January 30.

5. *POPLIFUGIA:* the Flight of the People, associated by the Romans with the retreat of a Roman army before the Fidenates at the time of the Gallic invasion in 390 B.C.; or with a panic of the people at the disappearance of Romulus in the Goat Swamp, an event usually connected with the Nones of this month. The term, Poplifugia, may have a ritualistic origin in lustrations performed to purify the people. Another interpretation makes this festival simply the celebration of the flight of an enemy. This is the only *fastus antiquissimus* occurring before the Nones of a month and there is some reason for thinking that the Poplifugia originally belonged to the cluster of *feriae* observed on the Nones of July. Cf. February 24.

   Feriae Iovi: rites to Jupiter, ordinarily on the Ides of a month. This extraordinary occurrence may be due to some association of Jupiter with the Poplifugia (see above), or to a decree of the senate in 42 B.C. ordering the celebration of Caesar's birthday on this day instead of on July 12.

6. *Ludi Apollinares:* games of Apollo instituted in accordance with the Sibylline books in 212 B.C. as a result of reversals in the war with Hannibal, and made an annual event in 208 B.C. after a pestilence. They were first celebrated on July 13 only and then extended to eight days, beginning on July 6. These games were supervised by the *praetor urbanus,* the city-praetor, and by the *decemviri sacris faciundis,* the ten men in charge of performing rites; they included both circus and theatrical events. Apollo was simply the Greek god as introduced from Etruria or from the Greek cities in Italy. At Rome he was sometimes identified with Soranus, the god of Mt. Soracte (cf. January 1 and November 13).

   Fortunae Muliebri: there were sacrifices at the fourth milestone on the Via Latina to the Good Fortune of Women. Only women of a single marriage could participate. Cf. April 1 and December 1.

7. *Nonae Caprotinae:* the Nones of the Goat when women, both slave and free, sacrificed to Juno Caprotina under a

caprificus or wild fig-tree on the Campus Martius. The juice of the fig-tree was used instead of milk in the sacrifice; and a twig or *virga* was cut from a goat-fig, perhaps for use in a fertility rite. Fig-milk was commonly used as a substitute for rennet in the making of cheese; and one of the best rennets was made from the stomachs of goats. That is why the so-called Ruminal fig-tree was planted near the shrine of Rumina at the Lupercal where Romulus and Remus were said to have been suckled by the she-wolf. A branch of the wild fig (male) was always used to fertilize the cultivated fig (female), an operation which occurs about this time of year in Mediterranean countries. A similar wedding of the date-palm may account for Palm Sunday in our Christian calendar. It is noteworthy that the rites of the Goat-Nones, sacred to Juno, do not occur in her month of June: this seems to be proof that the Roman festivals are in some cases older than the calendar, at least in its developed form. On one view, the Caprotine Nones originally fell in March, concurrent with rites for Vediovis (cf. January 1, March 7, and May 21), and marked the beginning of the 4-month year. It is thought that March was first called Caprotinus, and that the Nones took its name from this. Caprotinus has been interpreted as "Month of the Pig," then, by semantic change, "Month of the Goat."

Ancillarum feriae: rites of the servant girls, said to celebrate a deception of the Fidenates (cf. July 5) who demanded Roman wives and were duped into accepting servant girls dressed in the gowns of their mistresses. On this day serving women wore their mistresses' clothes and engaged in banter with one another. These rites may constitute the servile part of those described under the entry above. The story of the rape of the Sabine women (cf. August 21) may have originated in these rites of the *ancillae.*

Sacrificium Conso: a sacrifice to Consus, god of the stored harvest, at his underground altar in the Circus Maximus. More recently Consus has been interpreted as a god of copulation, and his position at this point in the calendar may derive from his functions as one deity of a breeding cycle which began on the Caprotine Nones. Cf. August 21 and December 15.

Palibus II: a festival of two Pales is marked in the Republican calendar for this date, but it does not appear in the later calendars of Imperial times. On one view, this is the original Palilia or Parilia, and the more famous one of April (cf. April 21) is only a copy of it. On this same view, the two Pales represent the twin kids of a prized she-goat, Caprotina; so in these rituals we may have the origin of the famous twin-legend, that of Romulus and Remus. Cf. April 21.

8. *Vitulatio:* the sacrifice of a *vitulus* or heifer, probably to Jupiter. Macrobius makes it a victory sacrifice connected with events subsequent to the Poplifugia (cf. July 5) which he dates on the Nones of July.

12. *Julius Caesar* was born on July 13, probably in 102 B.C.; but his birthday was celebrated the day before to avoid conflict with the last day of Apollo's games. Though belonging to an old patrician family, Caesar was associated from the beginning with the popular party; he married Cinna's daughter, and his aunt was the wife of Marius. Caesar's accord with Pompey and Crassus, the so-called first triumvirate, brought him the consulship in 59 B.C. and his subsequent command in Gaul. There he forged the army which later defeated Republican forces under Pompey at Pharsalus in 48 B.C. Caesar was dictator from 48 B.C. until his assassination by the liberators on the Ides of March, 44 B.C. His career put an end to the Roman republic and laid the foundations of monarchy and empire.

14. *Mercatus:* a 6-day fair or market following the *ludi Apollinares.* Cf. September 20 and November 18.

15. *Equitum probatio:* an equestrian review. Cf. September 14 and November 14.

18. *Dies ater:* anniversary of the Roman defeat by the Gauls on the banks of the river Allia in 390 B.C. This day was henceforth regarded as a *dies ater* or "black day." Cf. June 23 and October 6.

19. *LUCARIA:* rites said to have been celebrated in a *lucus* or grove where the Romans took refuge after their defeat by the Gauls at the river Allia (cf. July 18). The ceremonies probably date back to a much earlier period, but nothing is known of their real origin or nature.

20. *Ludi victoriae Caesaris:* circus games established by Julius Caesar in 46 B.C.

21. *LUCARIA:* see note on July 19.

22. *Concordiae:* anniversary of a temple to Concord, perhaps the one dedicated in the Forum by Camillus in 367 B.C. Cf. January 16.

23. *NEPTUNALIA:* rites of Neptune, the god of waters, at
first fresh waters. He was identified with an Etruscan
deity, Nethuns; with the Greek god, Poseidon Hippios;
and through him, with Consus. Neptune was paired with
a female deity, Salacia, whose name is derived from *salum,*
"salt water," from *salax,* "lustful," or perhaps from *salire,*
referring to leaping waters. Salacia was a name for Venus
as worshipped by prostitutes. Nothing is known about
the Neptunalia, except that celebrants protected them-
selves from the sun by huts made of foliage.

25. *FURRINALIA:* rites in honor of an unknown deity, Furrina. She has been explained as a native deity of spring-
water; or as an earth-mother, Furia, introduced from
Etruria and equivalent to the Greek Erinys or Fury. She
had a shrine on the slopes of the Janiculum hill near the
Pons Sublicius; and there was a *flamen Furrinalis.*

30. *Fortunae huiusque diei in campo:* anniversary of an unknown temple to the Fortune of this day.

AUGUST

1. *Spei ad forum holitorium:* a temple to Hope near the
Vegetable Market was dedicated during the first Punic
war and restored in Imperial times by Germanicus.

Marti Ultori: a temple to Mars the Avenger was dedicated
by Octavian either on August 1 or May 12. Cf. May 12.

Claudius, brother of Germanicus, was born on this day at
Lyons in France in 10 B.C. He succeeded his nephew,
Caligula, on the throne in 41 A.D. and ruled until his own
death in 54 A.D. Though never trained for his high position and more devoted to scholarship than to politics, he
reorganized the Roman government with great efficiency,
centralizing much of the administration in his own palace.
After the execution of his third wife, the infamous Mes-
salina, he married his niece, Agrippina the Younger, who
eventually poisoned him, set aside his children, and pro-
cured the throne for her own son, Nero. In Claudius' reign
Britain was first made a province, and the emperor himself
visited the British Isles in 43 A.D.

5. *Saluti in colle Quirinali:* the temple of Safety was dedicated in 302 B.C. during the Samnite wars. In 269 B.C.
its walls were painted by one of the consuls, Gaius Fabius,
whose descendants, including the historian Fabius, thus
bore the cognomen *pictor,* "painter." The temple was

struck by lightning four times and finally destroyed by fire in the reign of Claudius. As early as 180 B.C. Salus was identified with the Greek abstraction, Hygieia or Health.

9. *Soli Indigiti in colle Quirinali:* rites to the Native (?) Sun, held either on August 8 or 9, believed to be of Sabine origin. There was a *pulvinar Solis,* or couch of the sun, near Quirinus' temple, and the *gens Aurelia* or Aurelian clan had charge of the cult.

Pharsalus: Caesar defeated the Republican forces under Pompey in the decisive battle at Pharsalus in Thessaly on this day in 48 B.C.

12. *Herculi invicto ad circum maximum:* rites for the Unconquered Hercules near the Circus Maximus. Tithes were offered at the *ara maxima,* the great altar, and the Salii participated in ceremonies there. In historical times rites were also conducted by the *praetor urbanus* or city-praetor at the temple of Hercules in the Cattle Market: the praetor sacrificed a heifer, wore a laurel wreath, and kept his head uncovered in the Greek fashion. At the rites of Hercules no other god might be mentioned, no dog was admitted, and women were excluded. It was said that these rites had at one time been conducted by two patrician families, the Potitii and Pinarii, but that the latter were bribed by Appius Claudius, the censor of 312 B.C., to abandon their duties to public servants.

Hercules is doubtless just a Latinized form of the Greek Heracles. His cult was introduced at a very early date, since the *ara maxima* lies within the *pomoerium* or limits of the Palatine community. His rites were probably introduced by traders. For his identification with Semo Sancus Dius Fidius, see note on June 5. Vergil hands down a legend of Hercules' visit to the site of Rome and his conquest of the monster Cacus.

13. *Feriae Iovi:* the usual rites to Jupiter on the Ides of a month.

Dianae in Aventino: Servius Tullius, in conjunction with the Latins, was said to have dedicated a temple to Diana on the Aventine hill. The cult was of special interest to plebeians and slaves, particularly women, who washed their heads on this day. The temple of Diana may have been an asylum for fugitive slaves; it housed a primitive wooden statue of Diana. Diana is no doubt an old Italian goddess, as the root of her name appears in the names of

Jupiter, Dius, and perhaps Janus and Juno. As a goddess of light, of the groves, and of the chase, she was identified with the Greek Artemis as early as 398 B.C. As Lucina, she was a goddess of women in child-birth; and as Trivia, goddess of the Crossways, she was identified with the infernal Hecate. Her cult is said to have been brought from Aricia to Rome, perhaps when Rome assumed leadership of the Latin League. At Aricia she was associated with the nymph Egeria and with a male deity, Virbius, later identified with the Greek Hippolytus. Diana's rites at Aricia involved the strange slaying of a sacerdotal monarch (cf. February 24).

Vortumno in Aventino: a temple to Vortumnus or Vertumnus on the Aventine hill was dedicated in 264 B.C. He was a deity of fruits and gardens, linked with Pomona, the goddess of orchards, whose antiquity is proven by the existence of a *flamen Pomonalis.* There was a bronze statue of Vortumnus on the Vicus Tuscus, the Etruscan road, and the god was thought to be of Etruscan origin. This view is supported by the existence of a goddess Voltumna at Volsinii in Etruria. But Vortumnus may be of Latin origin, connected with *vertere,* "to change," and referring to the change of season, *annus vertens,* or to a turning back of the flooded Tiber, or to the changing nature of the god.

Herculi invicto ad portam Trigeminam: nothing is known about these rites to the unconquered Hercules at the Threefold Gate, but cf. August 12.

Castori Polluci in circo Flaminio: nothing is known about these rites for Castor and Pollux in the Circus Flaminius, but cf. January 27.

Florae ad circum maximum: nothing is known about these rites to Flora at the Circus Maximus, but cf. April 28.

17. *PORTUNALIA:* rites to Portunus at the Tiber port. This god seems to have had some connection with harbors, gates, and keys. He has been identified with the river-god Tiber and with Janus. There was a *flamen Portunalis* who anointed the arms of Quirinus.

Tiberinalia: Tiber-rites perhaps identical with the Portunalia; see above, and cf. December 8.

Iano ad theatrum Marcelli: a temple to Janus near the later site of Marcellus' theatre was vowed in 260 B.C. during the first Punic war. For another explanation of

Janus' connection with this day, see note under *Portunalia* above. Cf. January 11.

18. *Divo Iulio:* a temple to the deified Julius Caesar was dedicated on this day in 42 B.C. It became a conspicuous monument in the Forum.

19. *VINALIA:* the Vinalia Rustica or Vine Rites of the Country, a festival to celebrate the bringing of new wine into the city. It was either on this day or on April 23 that the *flamen Dialis* picked the first grapes and took auspices for the crop. Cf. April 23.

    Feriae Iovi: rites for Jupiter, perhaps connected with the role of his *flamen* in the Vinalia; see entry above.

    Veneri ad circum maximum: a temple of Venus was dedicated on this day in 295 B.C. and financed by fines imposed on matrons convicted of adultery. The chance dedication of temples to Venus both on April 23 and August 19 may explain her association with the vine.
    Augustus entered his first consulship on this day in 43 B.C., and he died on this day in 14 A.D. In honor of the first event, the month of Sextilis was renamed August after the prince.

21. *CONSUALIA:* rites of Consus (from *condere*, "to store"?) often regarded as god of the stored harvest. It is possible that he was at first a god of copulation connected with breeding cycles for animals (cf. July 7). The Consualia of August may have belonged to the Parilia of April (cf. April 21) in an original 4-month calendar. Consus had an ancient underground altar at the first goal-posts in the Circus Maximus; this is thought by some to mark the site of an ancient *penus* or store house. There was also a temple of Consus on the Aventine hill, dedicated in 272 B.C. At the Consualia there were horse races and mule races in the Circus Maximus; and a sacrifice was performed by the *flamen Quirinalis*, assisted by the Vestal virgins. The Rape of the Sabine Women was said to have occurred at this festival. The story may have been suggested by the rites of the *ancillae* on the Caprotine Nones (cf. July 7). Cf. December 12 and 15.

23. *VOLCANALIA:* rites of Vulcan, the fire-god, who had a temple of unknown age in the Circus Flaminius, as well as several consecrated areas in the city, including one in the *comitium* or assembly-place. On August 23 heads of families bought small fish at an *area Volcani* near the Tiber, and threw them into a fire. There was a *flamen Volcanalis* who sacrificed to Bona Dea on May 1, but his

participation in the Volcanalia is uncertain. Domitian erected an altar to Vulcan, at which a boar and a red heifer were sacrificed. The name Vulcan has been derived from an Etruscan gentilic name, Volca; and Vulcan has been regarded as an intruder, replacing Cacus, an earlier fire-god of the Palatine hill. The Volcanalia may have included sacrifice to Ops and Juturna. Cf. May 7 and 23.

24. *Mundus patet:* the *mundus* was a shallow pit on the Palatine hill, supposedly marking the center of Roma Quadrata, the earliest city. The *mundus* is thought to have been a *penus* or storehouse of the primitive city, or a trench dug at the foundation of this early city and filled with first-fruits. It was ordinarily covered with a *lapis manalis,* a "dripping stone" or a "ghost-stone." The *mundus* was opened on August 24, October 5, and November 8; and inhabitants of the underworld were thought to emerge from it.

25. *OPICONSIVIA:* the rites of Ops Consiva or Harvest Abundance, conducted at a shrine in the Regia which no one except the Vestals or a pontiff might enter. This shrine may have been the *penus* or store-room of the king's house. The name Ops shows the same root as *copia,* "abundance," and *optima,* "best." Ops was generally regarded as the wife of Saturn; but the cognomen Consiva has been interpreted to mean "wife of Consus." She was sometimes identified with Bona Dea (cf. May 1); and she received a sacrifice at the Volcanalia on August 23. Cf. December 19.

27. *VOLTURNALIA:* the rites of Volturnus, either a god of winds or a river-god. There was a river *Volturnus* in the Campania, and some believe that Volturnus was an old name for the Tiber. There was a *flamen Volturnalis* whose functions are nowhere described. The name Volturnus has been derived from that of an Etruscan family, the Volturia.

31. *Caligula,* the youngest son of Germanicus and the Elder Agrippina, was born on this day in 12 A.D. He was reared in his father's camp in Germany where the soldiers gave him this nickname, Caligula, or Little Boot, in reference to the diminutive soldier's shoes which he wore. When he grew up, he avoided his parents' quarrels with Tiberius; and ingratiating himself with the emperor at Capri, he managed to ascend the throne on Tiberius' death in 37 A.D. He proved to be a monstrous ruler, affecting divinity, insulting the senate, and pillaging the Empire until he was assassinated in 41 A.D.

1. *Iovi Tonanti:* a temple to Jupiter Tonans, Jove the Thunderer, was dedicated on the Capitol in 22 B.C.

   Iovi Libero, Iunoni Reginae in Aventino: a temple to Queen Juno was vowed by the dictator Camillus after the capture of Veii in 396 B.C. Augustus rebuilt this temple, and built or rebuilt one to Jupiter Liber (cf. March 17) on the Aventine hill.

2. *Actium:* Octavian and Agrippa defeated Antony and Cleopatra in this great naval battle off the coast of Greece in 31 B.C.

4. *Ludi Romani:* circus games celebrated probably in honor of Jupiter whose *epulum* or Banquet occurs in the middle of this period. Cf. September 13.

5. *Iovi Statori:* rites celebrating the founding of a temple to Jupiter Stator. Cf. January 13, June 27, and September 23.

9. *Aurelian* ruled from 270 to 275 A.D. and was largely occupied with resisting barbarian invasion and with suppressing Queen Zenobia of Palmyra. He is remembered chiefly for the wall which he built around Rome.

13. *Iovi epulum:* Jove's Banquet, probably a *lectisternium* or spreading of the couches for Jupiter, Juno, and Minerva on the birthday of the great Capitoline temple. The images of the gods were dressed for the occasion, and the face of Jupiter's image was painted red. His image reclined on a *pulvinar* or couch, while the images of the two goddesses sat on *sellae* or chairs. On the same day, for some religious reason, a *clavus* or nail was driven into the walls of Minerva's *cella* in the temple. Cf. November 13.

14. *Equorum probatio:* an equestrian review connected with the Ludi Romani. Cf. July 15 and November 15.

18. *Trajan,* born at Italica in Spain, was the first emperor of foreign birth. He succeeded Nerva in 98 A.D. and ruled until his own death in 117 A.D. Hailed officially and unofficially as *optimus princeps,* "the best of princes," he won the respect and devotion of his contemporaries. Both Pliny the Younger and Tacitus served under him as Roman officials. The column which he raised to celebrate his Dacian victories and which eventually contained his ashes, still stands as a memorial to the period of New Freedom established by his reign.

19. *Antoninus Pius* succeeded Hadrian as emperor in 138 A.D. and ruled until his own death in 161 A.D. Marcus Aurelius was his adopted son. Of provincial descent, Antoninus was known for his modesty and *pietas*. His reign was marked by good relations with the Senate and by further centralization of the Roman government.

20. *Mercatus:* a 4-day fair following the Ludi Romani. Cf. July 14 and November 18.

23. *Marti, Neptuno in campo; Apollini ad theatrum Marcelli:* anniversary of a temple to Mars in the Campus Martius, of one to Neptune in the same area, and of one to Apollo outside the Porta Carmentalis near the site of Marcellus' theatre.

Iovi Statori: rites marking the rededication of Jupiter Stator's temple. Cf. January 13, June 27, and September 5.

Iunoni Reginae ad circum Flaminium: rites marking the rededication of a temple to Queen Juno in the Circus Flaminius. Cf. September 1.

Augustus was born on this day in the consulship of Cicero and Antonius, i.e. 63 B.C., at the Ox-Heads, *ad Capita Bubula,* in the Palatine quarter of Rome. Born Gaius Octavius and later adopted by Julius Caesar in his will, he became Gaius Julius Caesar Octavianus. Caesar was his great-uncle. The title Augustus was conferred upon him in 27 B.C. after he had become sole ruler and established his Principate. His achievements are too vast to summarize here. We regard him as the first Roman emperor, but he represented his monarchy as a restoration of the Roman Republic.

26. *Veneri Genetrici:* a temple to Venus Genetrix, mother of Aeneas' race and particularly of the Julian clan, was dedicated by Julius Caesar in 46 B.C.

OCTOBER

1. *Tigillo sororio:* according to the Romans, this strange rite of "the sister's beam" was a relic of the famous combat between the Horatii and the Curiatii. The surviving Horatius, after slaying his sister for lamenting her lover's death, was ordered to make expiation by passing under a yoke represented by this beam which was let into opposite walls on a street. Under the beam, altars were erected to Janus Curiatius and to Juno Sororia. There was a Jupiter Tigillus or Beam at Umbrian Trebos.

Fidei in Capitolio: a sacrifice to Fides or Faith is ascribed by Livy to Numa. The *flamines* of Jupiter, Mars, and Quirinus rode in a covered chariot to the Capitol, their right hands wrapped in white cloth. According to Horace, Fides herself, i.e. her effigy, was veiled.

Alexander Severus, through the machinations of his mother, Mamaea, succeeded Elegabalus as emperor in 221 A.D. He ruled until his assassination in 235 A.D. Ulpian and Paulus, the jurists, were among his advisers. His death brought to an end the Afro-Syrian dynasty of the Severi under whom, collectively, Italy was reduced to provincial status.

3. *Ludi Augustales:* circus games established by Augustus on his return from Greece in 19 B.C., and later formalized by Tiberius.

4. *Ieiunium Cereris:* a fast to Ceres. Cf. April 19.

5. *Mundus patet:* see note on August 24; cf. also November 8.

6. *Dies ater:* a "black day" because the proconsul Caepio was defeated by the Cimbri on this day in 105 B.C. Marius retrieved Roman fortunes, and presently drove these Germanic hordes out of Italy. It was their first appearance south of the Alps. Cf. June 23 and July 18.

7. *Iovi fulguri:* rites to Jupiter of the lightning, or Jove the Flasher.

   Iunoni Curriti in Campo: rites to Juno Curritis or Quiritis (?), of Cures or of the *curis* or spear. Cf. February 1.

9. *Apollini in Palatio:* the temple to Apollo on the Palatine hill was dedicated by Augustus in 28 B.C.

10. *Iunoni Monetae:* rites marking the restoration of the temple of Juno Moneta. Cf. June 1.

11. *MEDITRINALIA:* on this day the new wine was tasted for the first time, along with a sip of the old, *medicamenti causa,* "to effect cures": hence the name of the festival, according to some. Celebrants repeated the words: "novum vetus vinum bibo, novo vetere (vino?) morbo medeor,"— "I drink the new wine and the old; I am cured of my illness (by the) new (wine) and (the) old." The goddess Meditrina may be named from the ritual.

12. *Augustalia:* rites celebrating the return of Augustus from the transmarine provinces. Cf. August 3.

13. *FONTINALIA:* a festival in honor of Fons, the deity of wells and springs which were garlanded on this day. There was a shrine of Fons on the Janiculum hill and a gateway of Fons in the Campus Martius.

15. *Feriae Iovi:* the customary rites to Jupiter on the Ides of a month.

October Horse: the right-hand horse of the winning team in a two-horse chariot race on the Campus Martius; it was slain and sacrificed to Mars: its tail was cut off and carried quickly to the Regia so that the blood would drip onto the sacred hearth or *focus.* Its head was also removed and carried off by the victors in a contest between the men of two districts, the Subura and the Via Sacra. If the former won, they fixed the head to the Turris Mamilia; if the latter won, they attached it to the wall of the Regia. The blood from the tail was mixed with the ashes of the unborn calves removed from the cows slaughtered at the Fordicidia on April 15; and this mixture was distributed to the people for some magical purpose at the Parilia on April 21.

Vergil was born at Andes near Mantua on this day in 70 B.C. when that part of Italy was still Cisalpine Gaul. He was educated in the Po valley during his childhood and came to Rome about 55 B.C. He was always attached to the cause of Caesar, and he may have served in the campaign which culminated in the defeat of Pompey at Pharsalus in 48 B.C. The *Bucolics,* published about 40 B.C., won Vergil immediate acclaim as a poet; but shy and sickly, he retired to the villa of a friend near Naples, and the rest of his life was spent, for the most part, in that area. The *Georgics* appeared about 30 B.C. and he was still working on the *Aeneid* at the time of his death in 19 B.C.

19. *ARMILUSTRIUM:* a purification of armor, probably by the Salii who, with the *ancilia* or sacred shields, made a procession on the Aventine hill near the Circus Maximus. It is thought by some that these rites originally marked the end of the campaigning season. Cf. March 23 and May 23.

25. *Ludi Victoriae Sullanae:* games established to celebrate Sulla's victory over the Samnites at the Colline Gate in 82 B.C.

28. *Isia:* festival of the Egyptian goddess, Isis, to whom Caligula erected a temple in the Campus Martius ca. 40 A.D. The worship of Isis involved mysterious rites including the use of "Nile water," music (especially that of the *sistrum* or rattle), dancing, and a dramatic ritual portraying the death and resurrection of Osiris, the husband of Isis.

NOVEMBER

1. *Ludi Victoriae Sullanae:* see note on October 25 when these games began.

   Isia: see note on October 28 when these festivities began.

4. *Ludi Plebeii:* these games took place in the Circus Flaminius which was dedicated in 220 B.C. They were celebrated by the plebeian aediles, and they may have been connected with the *epulum Iovis* of November 13.

6. *Agrippina the Younger,* daughter of Germanicus, was born in her father's camp at Cologne (later named for her, *Colonia Agrippinensis*). By Gnaeus Domitius Ahenobarbus she became the mother of Nero. Later she married her uncle, the emperor Claudius, and poisoned him to get the throne for Nero. Greedy, ambitious, and possessive, she became obnoxious to Nero himself and he had her murdered in 59 A.D.

8. *Mundus patet:* see note on August 24; cf. also October 5.

   Nerva was made emperor after the assassination of Domitian in 96 A.D. He was actually selected by the Senate, and inaugurated the era of New Freedom. He died two years later but left a worthy successor, Trajan.

13. *Feroniae in Campo:* probably a fair or market held in honor of Feronia, an ancient Italian goddess whose attributes are uncertain. She was worshipped by Roman freedwomen; and at Terracina her temple was the scene of manumissions. She was perhaps a deity of plenty, revered especially by plebeians descended from freed slaves. Her cult appeared in Rome before 217 B.C.; she had a temple in the Campus Martius. On Mt. Soracte she was associated with Soranus, perhaps as his wife and thus equivalent to Persephone: she was served by the *hirpi Sorani,* "the wolves of Soranus," a priesthood whose members walked on the burning embers of a pine-wood fire. Her name, like that of Fortuna, has been derived from *ferre,* "to bring," but it may have an Etruscan origin. Cf. January 1.

    Fortunae primigeniae in colle: a cult introduced from Praeneste in 204 B.C. The temple of Fortuna was dedicated ten years later. Cf. June 24.

    Epulum Iovi: see note on September 13.

14. *Equorum probatio:* an equestrian review connected with the Ludi Plebeii. Cf. July 15 and September 14.

16. *Tiberius* succeeded his step-father Augustus, on the latter's death in 14 A.D. A good soldier but sullen and indecisive on the throne, Tiberius had a long and undistinguished reign marked by mutinies in the armies, disagreements with his nephew, Germanicus, the treason of his minister, Sejanus, and the crucifixion of Christ. In disgust with politics he spent the last years of his rule in seclusion on the Isle of Capri, where he was murdered by his prefect, Macro, in 37 A.D. to secure the throne for Caligula.

17. *Vespasian,* commander of the Eastern armies at Nero's fall, was proclaimed emperor at Alexandria in Egypt in 69 A.D. Of equestrian stock, he founded a new dynasty and, under the title *imperator,* rehabilitated the Roman government. Simple and industrious, he set new examples in the moral sphere. He ruled until his death in 79 A.D. when he was succeeded by his eldest son, Titus.

18. *Mercatus:* a 3-day fair succeeding the Ludi Plebeii. Cf. July 14 and September 20.

DECEMBER

1. *Neptuno:* anniversary rites for founding of a temple to the god of waters. Cf. July 23.

   Pietati: a temple to Pietas or Piety was dedicated in 181 B.C.

   Fortunae Muliebri: rites celebrating the establishment of an altar to Womanly Fortune. Cf. April 1 and July 6.

3. *Sacra Bonae Deae:* these rites of the Good Goddess were held in the Regia or in the adjacent house of the Pontifex Maximus, and celebrated only by women under direction of the Pontiff's wife. It was at this festival that Cicero's enemy, the infamous Clodius, was caught in women's attire in 62 B.C. Caesar, the Pontifex Maximus, divorced his wife, and a sensational trial of the culprit ensued. Cf. May 1.

5. *Faunalia rustica:* rural rites of the god Faunus, a woodland deity identified with the Greek Pan. His name has been derived from *favere* and *fari,* suggesting that he was a propitious deity with prophetic powers. Vergil lists him as the third king of Latium. His name is often used in the plural. These rites were perhaps originally the termination of a breeding cycle for goats. Cf. February 13.

8. *Tiberino in insula:* rites for Father Tiber on the island in his stream. Vergil makes him a river-god who appears to

Aeneas before the latter's visit to the site of Rome. Cf. August 17.

11. *AGONALIA:* see note on January 9; cf. March 17, and May 21. The *agonalia* of December 11 were definitely held in honor of one or more *indigetes,* and they have been interpreted as the termination of a breeding cycle for goats which began on the Caprotine Nones in a 4-month year. In that primitive era, they would have fallen in April. They may be connected with the Septimontium (see below) which, like the Caprotine Nones and the Parilia, appears to be another festival for Pales.

Septimontium: a festival of the *montani* or hill-men who inhabited the seven hills of the early city. These rites were conducted by the *Flamen Palatualis,* so they may have been sacred to Pales. On this day no beast-drawn vehicle was permitted within the city. Cf. April 21 and July 7.

12. *Conso in Aventino:* rites honoring Consus, the god of the grain-storage. Cf. August 21.

13. *Telluri et Cereri in Carinis:* rites celebrating the establishment of a temple to Earth and Ceres in the Keels quarter of Rome.

15. *CONSUALIA:* the winter rites of Consus which, like those of August 21, took place at his underground altar in the Circus Maximus. Horses and asses were adorned with flowers; and there were mule races in the Circus Maximus. Cf. August 21.

Nero, son of Agrippina the Younger and Gnaeus Domitius Ahenobarbus, was born on this day in 37 A.D. As a youth he was tutored by the philosopher Seneca who later became his political adviser. An aesthete rather than a statesman, Nero succeeded Claudius on the throne in 54 A.D.; and his violent and disastrous reign was marked by the slaughter of intimate friends and relations, as well as political opponents. The great fire swept Rome in 64 A.D.; and Nero, looking for scape-goats, accused the Christians of arson and indulged in the first major persecution of that sect. When his armies revolted in 68 A.D., Nero committed suicide and brought to an end the Julio-Claudian dynasty.

17. *SATURNALIA:* the festival of Saturn whose name is usually derived from *serere,* "to sow," suggesting that he was a god of sowing and perhaps of the harvest, since he carried a sickle. Recent interpretation derives his name from Satre or Satria, an Etruscan family name. Vergil

represents him as the first king of Latium and ruler of the
Golden Age in Italy. He was identified with the Greek
Titan Cronus who fled the Olympians and found a hiding-
place in Latium (as if from *latere*, "to hide"). He was
always worshipped in the Greek fashion, with head uncovered. He was associated with the Capitoline hill and
a Saturnian settlement there; and his temple in the Forum
just at the base of the Capitoline became the *aerarium* or
treasury of the Roman Republic. Saturn was closely associated with Janus in Roman myth. Ops or Plenty was
usually regarded as Saturn's wife, but Gellius gives his
cult partner as Lua, a power of blight and a deity who
expiated blood shed in battle.

The Saturnalia lasted from three to seven days in historical times. On December 17 there was a public sacrifice
and feast; the following days were devoted to household
festivities and gift-giving. Wax tapers, *cerei*, and clay
images, *sigillaria*, were the customary gifts. Slaves were
allowed to wear the *pilleus* or freedom-cap; they were
served by their masters at dinner and in general permitted to enjoy "December's liberty." There was a prince
of the Saturnalia whose rule expired with the holidays;
he may represent a sacerdotal monarch of very early
times who was expelled or even slain at the expiration of
his unusual office which perhaps involved supervision of
an intercalary period. Cf. February 24.

19. *OPALIA:* a festival of Ops or Plenty. Traditionally, she
was the wife of Saturn, and her festival got closely associated with the Saturnalia. She seems to be paired with
Consus, however, in the festival of the Opiconsivia on
August 25.

21. *DIVALIA:* rites of the goddess Angerona about whom
little is known. The Elder Pliny states that her cult statue
had its mouth bound and sealed. She was worshipped in a
place called the *sacellum Volupiae*, which may mean the
"chapel of Pleasure." But this phrase has been more
recently interpreted as "chapel of the wolf-goddess." In
antiquity Angerona's name was derived from *angina*,
"quinsy," or from *angor*, "anguish." Modern explanations
include a derivation from obsolete *angerere*, "to raise up,"
referring to the sun after the winter solstice; derivation
from an Etruscan family name; derivation from obsolete
angus as in *angustiae*, suggesting that she was a goddess
of the Narrows, i.e. the jaws of Hell; and derivation from
the root found in *annus* and *agonium*, indicating that
Angerona was a yearling deity. In a 4-month year the
Divalia would have been concurrent with the Parilia of

April (cf. April 21) and with the Consualia of August (cf. August 21). The Divalia, like the Parilia, seem to have had some relation to the beginning of a "year" in the sense of a breeding cycle. Cf. also December 23.

22. *Laribus permarinis:* a temple to the Lares who travel with seafarers was dedicated in 179 B.C.

23. *LARENTALIA:* funeral rites at the grave of Acca Larentia, wife of Faustulus and foster-mother of Romulus and Remus. Faustulus is sometimes regarded as another form of Faunus. Acca Larentia was sometimes represented as the mother of the Lares, and also of the original Arval Brothers. The successors of the latter, twelve in number, constituted a priesthood of the *arva* or ploughed fields: they worshipped Mars, the Semones, and the Lares in the grove of Dea Dia at the fifth milestone on the Via Campana outside the Porta Portuensis. The Larentalia were conducted by the *Flamen Quirinalis;* they may constitute a dislodged portion of the Divalia—in this case, Acca's name is probably a by-form of Angerona (cf. December 21).

Feriae Iovi: rites to Jupiter, occurring ordinarily on the Ides of a month. On this occasion, they may have been connected with the Larentalia.

25. *Natalis invicti solis:* the birthday of the Unconquered Sun, celebrated by the followers of Mithras, an Indo-European deity associated with Ahura-Mazda, the power of goodness and light in the Persian religion of Zoroaster. The cult of Mithras is said to have been introduced into the West by the pirates whom Pompey suppressed in 67 B.C. The rites of Mithras included a sacramental meal, an initiation of seven grades including the status of *miles* or "soldier," a baptism, and worship of Mithras with his sacred bull. This cult was very popular with Roman soldiers whose influence, upon conversion to Christianity, helped to consecrate the birthday of the Unconquered Sun as the birthday of Christ. There is a similar significance in the consecration of Sunday as the Lord's Day and in the fixing of Easter on a Sunday. The birthday of the Sun is not listed in the calendars of the early Empire but it appears in the calendar of Philocalus (354 A.D.).

30. *Titus,* eldest son of Vespasian, succeeded his father in 79 A.D. and ruled only until his untimely death in 81 A.D. His reign is memorable for the eruption of Vesuvius in 79 A.D. He completed the Colosseum begun by his father; and the Arch of Titus celebrates his victories over the Jews. He was succeeded at his death by his tyrannical brother, Domitian.

V. GLOSSARY OF GODS

This list includes all deities mentioned in the notes to the calendar, with the exception of most Greek gods and obvious abstractions. Greek gods without well-known Roman equivalents, e.g. Apollo and Hercules, are included. Gods are listed as separate deities when they may have originated in the identification of separate deities. The calendar dates given below (by month and day) are all those under which the name of a god is mentioned in the notes; and the date for fullest reference is often in **bold face.**

ACCA LARENTIA   Jan. 3; Feb. 17; **Dec. 23**

AESCULAPIUS (ASCLEPIUS)   Jan. 1

ANGERONA   Dec. 21; Dec. 23

ANNA PERENNA   Mar. 1, **15**; June 18

APOLLO   Jan 1; **July 6**; Sept. 23; Oct. 9

ATTIS   Mar. 25

BELLONA   Mar. 1, 19; June 3

BONA DEA   **May 1**, 15, 23; June 11; Aug. 23, 25; Dec. 3

CACA   May 7; June 9

CACUS   May 7, 23; **Aug. 23**

CARDEA   June 1

CARMENTA (CARMENTIS)   **Jan. 11**, 15

CARNA   June 1

CASTOR, CASTORES (see DIOSCURI)

CERES   Jan. 24; Mar. 15; Apr. 12, **19**; May 29; Oct. 4; Dec. 13

CONSUS   Feb. 17; Mar. 17; Apr. 21; July 7; **Aug. 21**, 25; Dec. 12, 15, 19

CYBELE   **Mar. 25**, 27; Apr. 4, 10

DEA DIA   Dec. 23

DIANA   Jan. 11; Feb. 24; Mar. 31; Apr. 24; **Aug. 13**

DIESPITER (see JUPITER)

DIOSCURI (CASTORES or CASTOR and POLLUX) Jan. 11, **27**; Aug. 13

DIS PATER May 15

DIUS FIDIUS (see SEMO SANCUS)

EGERIA Aug. 13

FATUCLUS Feb. 13

FATUUS Feb. 13

FAUNA Feb. 13; May 1

FAUNUS Jan 1; **Feb. 13**, 15; Apr. 28; Dec. 5, 23

FERONIA Jan. 1; Nov. 13

FLORA **Apr. 28**; Aug. 13

FONS Oct. 13

FORNAX Feb. 17

FORS (see FORTUNA)

FORTUNA Apr. 1, 5; May 25; June 11, **24**; July 6, 30; Nov. 13; **Dec. 1**

FURIA July 25

FURRINA July 25

HERCULES June 4, 5, 30; **Aug. 12**, 13

INCUBO Feb. 13

INDIGES **Jan. 9**; Mar. 17; Aug. 9; Dec. 11

INUUS Jan. 9; Feb. 13, 15; Mar. 17

ISIS Oct. 28

JANUS Jan. 7, 9, **11**; Mar. 30; Apr. 24; Aug. 13, 17; Oct. 1; Dec. 17

JUNO **Feb. 1**, 15; Mar. 1; Apr. 24; June 1; July 7; Aug. 13; Sept. 1, 13, 23; Oct. 1, 7, 10

JUPITER Jan. 1, 5, 11, 13; Feb. 17, 23; Mar. 17; Apr. 1, 13, 23, **24**; May 15; June 5, 9, 13, 20, 27; July 5, 8; Aug. 13, 19; Sept. 1, 4, 5, 13, 23; Oct. 1, 7, 15; Nov. 13; Dec. 23

JUTURNA **Jan. 11**; May 23; Aug. 23

LARA Jan. 3

LARES **Jan. 3**; Feb. 22; May 1; June 27; Dec. 22, 23

LEMURES May 9

LIBER   Feb. 15; **Mar. 17**; Apr. 19; Sept. 1

LIBERA   Apr. 19

LUA   Dec. 17

LUNA   (see DIANA)

MA   June 3

MAGNA MATER   (see CYBELE)

MAIA   May 1, **15**, 23

MAIESTA   (see MAIA)

MAMURIUS VETURIUS   Mar. 1, **14**

MANES   Feb. 21; May 9

MANIA   Jan. 3

MARS (MAVORS, MARMAR, etc.)   Feb. 17, 27; **Mar. 1, 9,**
   14, 15, 19; Apr. 1, 24, 25; May 12, 14, 29; June 1; Aug. 1;
   Sept. 23; Oct. 1, 15; Dec. 23

MATER MATUTA   June 11

MEDITRINA   Oct. 11

MERCURY   May 15

MINERVA   Feb. 1; Mar. 1, 15, **19**; June 13, 19; Sept. 13

MULCIBER   (see VULCAN)

NEPTUNE   **July 23**; Sept. 23; Dec. 1

NERIO (NERINE)   Mar. 1, 19; Apr. 1; June 3

NETHUNS   July 23

NORTIA   Apr. 1; June 24

OPS   May 1; Aug. 23, **25**; Dec. 17, 19

OSIRIS   Oct. 28

PALES   Apr. 21; July 7; Dec. 11

PENATES   May 7

PERNA   Mar. 15

PLUTO   (see DIS PATER)

POLLUX   (see DIOSCURI)

POMONA   Aug. 13

PORRIMA (PRORSA)   Jan. 11

PORTUNUS   Aug. 17

POSTVERTA   Jan. 11

QUIRINUS   **Feb. 17;** Mar. 1; Apr. 24; June 29; Oct. 1

ROBIGUS   Feb. 17; **Apr. 25**

SALACIA   July 23

SATURN   May 15; Aug. 25; **Dec. 17,** 19

SEMO SANCUS (DIUS FIDIUS)   June 4, **5;** Aug. 12, 13

SOL   Aug. 9; Dec. 25

SORANUS   Jan. 1; July 6; Nov. 13

SUMMANUS   June 20

TELLUS   Jan. 24; Apr. 15, 19; May 15; Dec. 13

TERMINUS   Feb. 23

TIBERINUS   May 15; Aug. 17; Dec. 8

TRIVIA   Aug. 13

VACUNA   June 3

VEDIOVIS (VEIOVIS or VEDIUS)   **Jan. 1;** Mar. 7; May 21; July 7

VENUS   Mar. 1; **Apr. 1,** 23; July 23; Aug. 19; Sept. 26

VESTA   Mar. 1; May 7; June 7, **9,** 15

VICA POTA   Jan. 5

VIRBIUS   Aug. 13

VOLTUMNA   Aug. 13

VOLTURNUS   Aug. 27

VOLUPIA   Dec. 21

VORTUMNUS   Aug. 13

VULCAN   May 7, 15, **23;** Aug. 23

VI. BIBLIOGRAPHY

Principal works for study of the Roman calendar include:

Sources

Censorinus, *de Die Natali*, an astrological work of the third century A.D., but reliable in its borrowings from Varro (see below).

Corpus Inscriptionum Latinarum, vol. 1 (second edition): calendars of imperial date, as edited by Mommsen.

Inscriptiones Italiae, vol. 13, fascicle 2 (1963): the first complete collection of ancient Roman calendars, edited by A. Degrassi, and handsomely published by the Italian government.

Livy, *Libri*, the history of Rome from its founding, by the great historian of the Augustan age.

Macrobius, *Saturnalia*, a valuable miscellany of the fourth or fifth century A.D.; the first book contains our best account of the origin and development of the Roman calendar.

Notizie degli Scavi di Antichita (1921), pp. 10-141; Mancini's publication of the Fasti of Antium, the only extant calendar antedating Caesar's reforms.

Ovid, *Fasti*, a long poem by the brilliant Augustan poet, dealing with the festivals and anniversaries of the first six months of the year.

Plutarch, *Quaestiones Romanae*, an aetiological study by the learned Greek polymath and biographer of the first and second century A.D.

Varro, *de Lingua Latina*, a work on etymology and semantics by the scholarly friend of Cicero.

Vergil, *Aeneis*, Rome's epic preserving many facts of religious interest, especially in Book 8.

Commentaries

Altheim, F., *History of Roman Religion* (New York, 1937) : the author finds Etruscan origins for many Roman institutions and beliefs.

Colson, F. H., *The Week* (Cambridge, 1926) : to date the most thoughtful and complete essay on this subject.

Fowler, W., *Roman Festivals* (London, 1899) : still the most complete handbook on the subject in English, but out of date in some respects.

Frazer, J. G., *Fasti of Ovid* (London, 1929) : a monumental work with copious notes by the famous author of the *Golden Bough:* his analogies are world-wide in scope.

Latte, K., *Roemische Religionsgeschichte* (Munich, 1960) : the latest handbook on Roman religion.

Michels, A. K., *Calendar of the Roman Republic* (Princeton, 1967) : a close examination of the pre-Julian calendar.

Neugebauer, O., *Exact Sciences in Antiquity* (Providence, 1957) : now an indispensable guide for the study of ancient astronomy and mathematics.

Nilsson, M. P., *Primitive Time Reckoning* (Lund, 1920) : the classic work on this subject.

Odom, R. L., *Sunday in Roman Paganism* (Washington, 1944) : written from a particular religious viewpoint, but containing a good array of evidence for the planetary week.

Rose, H. J., *Ancient Roman Religion* (London, 1948) : a very instructive and easy account by an acknowledged authority; Professor Rose is the author of numerous pertinent articles in the *Oxford Classical Dictionary* (Oxford, 1949).

Rose, H. J., *Handbook of Greek Mythology* (London, 1958) : a concise account of Greek myths and the extension of them to Roman subjects.

Rose, H. J., *Primitive Culture in Italy* (London, 1926) : in this work the author first outlined his famous views on *sacer* as "tabu" and *numen* as "mana."